STUDY GUIDE TO HUMAN ANATOMY AND PHYSIOLOGY 2

By Michael Harrell, M.S.

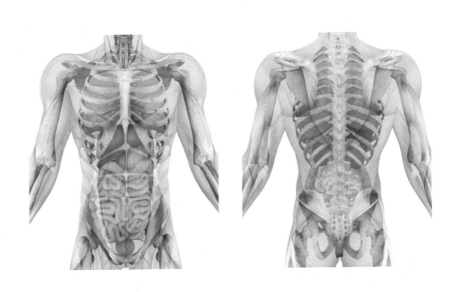

© 2012 Professor Michael Harrell, All rights reserved.

No part of this book may be reproduced, stored, or transmitted by any means without the written permission of the author.

Published September, 2012

Dedication

To my wife for her love, dedication and support.

Table of Contents

PREFACE	5
HOW TO LEARN HUMAN ANATOMY AND PHYSIOLOGY	7
ENDOCRINE SYSTEM	12
BLOOD	54
HEART	83
ARTERIES AND VEINS	113
LYMPHATIC SYSTEM	143
RESPIRATORY SYSTEM	166
DIGESTIVE SYSTEM	196
URINARY SYSTEM	219
WATER, ELECTROLYTES AND ACIDS	244
REPRODUCTIVE SYSTEM	256
DEVELOPMENT	273
APPENDIX	286

PREFACE

Welcome everyone to your guide to Human Anatomy & Physiology! I have been teaching college level human anatomy and physiology for many years, as well as other courses. My other classes taught have included: pathophysiology, biology, zoology, microbiology, and others. In this time I have seen thousands of students. I have learned through the years the best ways to learn the most information in the least amount of time. There are two ways to study, smart or hard. If you will follow my information and learn the key points of each chapter, you will make an excellent grade in your A&P class. In each chapter concentrate your efforts on learning the key terms. The key terms are the ones you are most likely to see on your exams. Learn to associate words and how to connect them.

For example, "Anatomy is the study of the structure of the human body." Look at the key words in this sentence, anatomy and structure. Learn how to pick out these key terms and remember them, not the entire sentence or paragraph full of information. When given a paragraph, page or whatever; just memorize the key words and then learn how to associate them. Learn what they have in common and be able to speak from one word to the next. This will be the best way to learn your anatomy text.

I will make the assumption that anyone reading this book is taking human anatomy and physiology. You will still need your text, but more as a reference to pictures and such. This guide will give you the important information from the chapters, which will be what you are most likely to see on an exam. Sample questions will be included, which are also the most likely for you to see on an exam.

Note also that this book is not a guide for A&P lab. An anatomy lab book is little more than a book with lots of pictures in it. That is what anatomy is, memorizing parts and pieces of the body. You simply look at the picture in your book and then learn those parts on a model. You may be looking at a skull, brain, kidney, etc., it is simple memorization. This book is more to help you with the lecture.

HOW TO LEARN HUMAN ANATOMY AND PHYSIOLOGY

Here's how you're going to learn the information contained in your anatomy and physiology text. First of all, an enormous amount of information is going to come at you in a very short amount of time. Most people spend their lives avoiding science, so you probably won't be prepared. Consider how much time you would usually spend studying for a class and triple that amount. If you want to learn to apply what is ahead of you, then now is the time to start working.

Points to remember:

1. A great attitude can take you far. Any instructor prefers a C student with a good attitude over an A student with a bad attitude any day of the week. No matter how tough things get, never give your instructor a hard time. That instructor has one big power and

that is the grade book. Don't forget this because I can assure you, your instructor won't.

2. Develop good study habits and don't let anything disrupt them. Human are creatures of habit, so if you get into a regular studying routine, you will probably stick with it.

3. Get rid of anything which will be a distraction to you, when you are studying. Ask yourself what is it that you always allow to draw your attention, while you are studying? Find isolation during your set time to study and don't allow anything to be close enough to you, which would cause a disruption.

4. When studying, read your notes or other study material our loud. If you are listening to the information at the same time you are reading it, you are more likely to remember it. People may think you are crazy, but they will be the same people, who are failing the class, so don't worry about it.

5. Record yourself reading your notes with some type of audio device, so you can play this back to yourself, when you have time. Many students have long drives to school and this is time you could

be listening to something helpful. Don't turn on the radio or listen to useless music, when you could be studying. Most people spend at least an hour or two in the car each day, so use this time wisely. How would you like to have an extra two hours learning every day?

6. After you feel like you understand the information, explain it to someone else. If you can thoroughly explain a topic to someone, by using complete sentences and not saying, "Ummmm", then you might just know what you are doing.

7. Make yourself a short study sheet and put it in your pocket. Whenever you have a few minutes, in between class or whatever, pull out the paper and look over it. Those minutes add up over days.

8. Draw pictures of your topics whenever possible. If you force yourself to make an illustration of structures, cells, etc., it will be easier to remember them. Try it and you will find it helps.

9. When you start to feel comfortable on the next set of test material, make up your own test. You think it's easy to make out a 100 question test? Just try it and you will see it isn't so easy. After writing your own questions, give them to a friend. When they start

to tell you that they don't understand the question, you will discover it isn't as easy as you think.

10. Now this one is important people, so don't blow it! The greatest skill you can ever learn is how to listen. What is that you say? You know how to listen? Have you ever just tuned someone out? Like maybe your teacher in your last class? You can hear someone speaking, but that doesn't mean you are hearing the words. If you can't repeat exactly what the speaker just said, then you weren't listening. Most people will never master this skill and their grades will reflect it.

11. College is not something you do in your spare time. When you start college, it becomes the most important thing you have to do and life is secondary. If you can't find the time to put college first, then you may not succeed.

What you will find in this book.

The quantity of information contained in the average human anatomy and physiology text is enormous and the average person can't absorb that much information in a short period of time. What you will want to do is to learn key words and phrases for a block of information. That among other learning exercises will help you to retain information quicker and more efficiently.

In example, imagine that you just saw a movie and wanted to tell a friend about what you saw. Obviously you couldn't remember everything that happened in the movie, but you would remember things such as: key characters, important events, etc. When you have a large volume of information, what you want to do is to remember key words and know how to connect them. In other words, pick out the important words and know what they have in common. If you can connect the words by telling a story, you will probably have a reasonable grasp of the information.

If you look at the organization of any human anatomy and physiology text, they are all organized in the same manner. You will first find an overview of the book, then a section on chemistry, then the cell, etc. We will cover those topics in the same order and learn them one at a time. There is no sense in fooling around any longer, let's get started.

ENDOCRINE SYSTEM

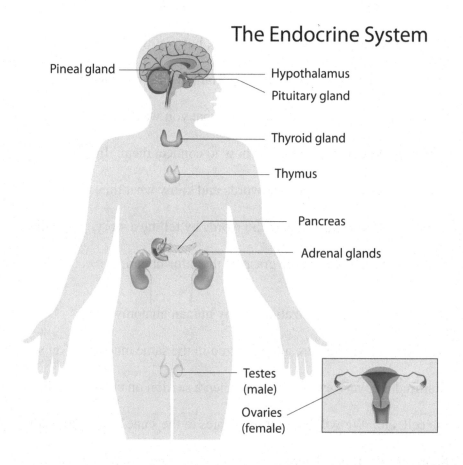

CHAPTER 1
ENDOCRINE SYSTEM

The endocrine system is all about one thing, hormones. Anytime you hear hormones, you should always think "chemical signals." A hormone is by definition a chemical signal which enters the blood and travels some distance to its target tissue. When you think of chemical signals, think about ligands. Remember ligand gated ion channels are protein channels which open or close in response to a chemical signal. Hormones work on cells in three major ways and affecting ion channels is the number one way they work.

Consider drugs and how they work on our bodies. Many drugs we take work to mimic or block chemical signals; we already have in our body. Many of these drugs are opening or closing ion channels. After you learn how the major hormones of the body work, it is often easy to understand how many drugs work. Many drugs mimic or block a chemical signal; we already have in our body. This information on hormones will be very useful, if you take pharmacology.

The major endocrine organs of the body are: hypothalamus, pineal gland, pituitary gland, thyroid, parathyroids, thymus, adrenals, pancreas, ovaries and testes. Notice how some of the endocrine organs also belong to other systems. Before we look at each one of these organs, let's look at the endocrine system itself.

Hormones work on one or more groups of cells called target tissues. Some hormones may work on one particular cell type and other hormones may work on many cell types. Students often ask, "How do hormones know where to go in the body, so they affect one particular cell type?" That is not how hormones work. Remember hormones enter the blood and after they enter the blood, they are transported to almost all tissues of the body. Some cells will have the receptors for a hormone and other cells won't.

Remember that all cells don't exhibit the same proteins. Look at the proteins in the plasma membrane, many of these proteins are found on most cells, but some proteins are found on only one cell type. So, when a hormone enters the blood, it will travel to most tissues, but it will only affect the cells which have the protein specific to that hormone.

Hormones can be classified in many ways; a functional classification would be as follows:

1. autocrine chemical signals – chemicals which act locally and only on the same cell type which secreted them. These chemicals work on cells close to the secreting cells and only on the same type which secreted them.

2. paracrine chemical signals – paracrine signals also act locally, but on a different cell type than the secreting cells.

3. hormone – unlike the first two, hormones don't act locally. Hormones enter the blood and travel some distance to the target tissue. Most of the chemicals we will look at in this chapter will be of the hormone classification. Most hormone secreting cells are epithelial cells, but some are neurons secreting neurohormones.

4. neurohormones – chemical signals secreted by neurons, which enter the blood and travel some distance to target tissues.

5. neurotransmitter – chemical secreted between two neurons. These chemicals will be seen crossing synapses from one neuron to another.

6. pheromones – chemicals used to affect individuals around the individual secreting it. These chemicals aren't used to affect the producing individual, but they do affect the physiology and behavior of individuals in the environment.

The endocrine system and the nervous system are the two systems of the body which have control over the body. In many ways these two systems can't be separated and the two systems are often discussed together. When you look at the organs of these two systems, you will notice that some of the organs of one system overlap into the other.

Even though the two systems are connected and work together, they work in different ways. The nervous system uses electric signals (action potentials) and chemical signals (hormones, ligands) to control the body. The electric signals work by the all or none principle, which means the electric signals are of the same strength. Because of this the central nervous system must monitor the frequency of action potentials. Our brain monitors our blood pressure by determining the frequency of action potentials returning from baroreceptors in some major arteries. Low frequency means low blood pressure and a high frequency means high blood pressure. In addition, the nervous system also uses many chemical signals.

The endocrine system uses chemical signals only. The quantity of the chemical signal is what determines the strength of the response. This is easily seen in an example. If someone consumes a small amount of a chemical, such as alcohol, the person sees a small change in the body. If a person consumes a large amount of alcohol they see a big change in the body. So quantity is very important when it comes to the strength of the stimulus with hormones.

Water soluble vs. lipid soluble hormones

Hormones work on cells in three possible ways: open/close ion channels, initiate DNA synthesis or activate enzymes. When you consider where a water vs. lipid soluble hormone will work, think about what the plasma membrane is made of. The plasma membrane is made mostly of two materials: lipids and proteins. What will happen when a water soluble material comes in contact with a lipid material? Everyone knows water and oil don't mix. So a water based hormone will not diffuse through the lipids of the plasma membrane. Because a water based hormone can't move through the membrane, it must work in the extracellular environment. When hormones work in the extracellular environment they usually work on ion channels. Opening and closing ion channels are how you will usually see water based chemical signals working. Water soluble hormones may also activate enzymes.

Lipid soluble hormones will diffuse through the cell membrane and work on the inside of the cell. The lipid soluble chemicals can diffuse through the cell membrane, since the cell membrane is half lipids. Where water and lipids won't mix, lipids and lipids will mix. The lipid soluble hormones often work by initiating DNA synthesis and of course the DNA is on the inside of the cell. Lipid soluble hormones may also activate enzymes.

HORMONE REGULATION

Most hormones are regulated by a negative feedback mechanism. Remember negative feedback is when the body works against a deviation from homeostasis. Occasionally a positive feedback mechanism will control a hormone.

Hormones may also be regulated by the nervous system. Remember that some organs of the endocrine system are also located within the nervous system. The sympathetic division of the nervous system releases norepinephrine and epinephrine from the adrenal medulla. This example illustrates how the nervous system can regulate hormone release.

Hormones may also be regulated by tropic hormones. A tropic hormone is a hormone which regulates the release of another hormone. Growth hormone releasing hormone from the hypothalamus regulates the release of growth hormone from the anterior pituitary. This illustrates one hormone regulating the release of another.

Substances other than hormones can also regulate the release of a hormone. Insulin is regulated by blood sugar levels. When we eat a meal, our blood sugar levels will rise. As blood sugar levels rise, insulin levels will rise. Insulin will cause most cells of the body to absorb sugar and this will lower blood sugar levels.

HORMONE RELEASE PATTERNS

Some hormones are released in very regular patterns. Think about the hormones involved with ovulation and the menstrual cycle. Estrogen and progesterone follow regular patterns unless something like pregnancy occurs.

Some hormones can greatly increase in the blood over a short period of time. When the sympathetic division of the nervous system is activated, we see rapid rises in epinephrine levels. Not only will this hormone rapidly rise, but it will also rapidly decrease. When we are startled we see a rapid increase in heart rate, but it doesn't stay elevated for long.

Other hormones will stay at very constant levels. Thyroid hormone stays within a very narrow range and will only show significant changes over long periods. If someone wishes to elevate their metabolism, they can't exercise for only a few weeks. They must put increased energy demands on the body for many months and years, before a substantial increase in thyroid levels are seen.

ORGANS OF THE ENDOCRINE SYSTEM

Hypothalamus

The hypothalamus is found in the brain, superior to the pituitary gland. These two structures are found close to each other and have two very special connections to each other. The hypothalamus controls the two halves of the pituitary through two separate systems. Let's look at these two connections.

They hypothalamus connects to the posterior pituitary through the infundibulum creating a connection called the hypothalamohypophysial tract. When you think of this tract, think of axons, because that is what travels through the tract. The posterior pituitary is nothing more than an extension of the hypothalamus. Imagine if you could grab the axons of the neurons in the hypothalamus and pull them down. If the axons would stretch down, then the ends of these axons would be the posterior pituitary.

The posterior pituitary is also called the neurohypophysis because it's an extension of the hypothalamus, which is part of the nervous system. So when you think of the posterior pituitary, think nervous system, because it is part of it. The hormones you see released from the posterior pituitary are all produced in the hypothalamus, so they will be categorized as neurohormones. Neurohormones are hormones produced by a neuron and released into the blood. Since the posterior pituitary is made up of the terminal ends of the axons, any hormones released from them must have come from the cell bodies of the neurons which produced them and those cell bodies are located in the hypothalamus.

The anterior pituitary is also called the adenohypophysis and is not part of the nervous system, like the posterior pituitary is. The anterior pituitary is composed of epithelial cells. These epithelial cells come from the oral cavity, far back in embryonic development. The hypothalamus connects to the anterior pituitary through the hypothalamohypophysial portal system.

Several portal systems will be discussed in anatomy and physiology. A portal system will always be two sets of capillary networks, with a direct connection in between. The body sets up portal systems, so that it can deliver materials from one point to another directly. Imagine two points with a one way road in between them. Whatever leaves point A can only go to point B and nowhere else.

They hypothalamohypophysial portal system is in the body so that anything entering the portal system from the hypothalamus can only go to the anterior pituitary. This system gives a very precise delivery of hormones. With hormones a little bit goes a long way, so precise delivery is important to maintaining homeostasis. The hypothalamus regulates hormone release from the anterior pituitary by releasing tropic hormones through the portal system. A tropic hormone is a hormone which controls the release of another hormone.

The hormones leaving the hypothalamus and traveling through the portal system to the anterior pituitary are:

1. prolactin-releasing hormone (PRH) – this hormone will travel from the hypothalamus, through the portal and cause more prolactin to be released from the anterior pituitary.

2. prolactin-inhibiting hormone (PIH) – this hormone slows the release of prolactin from the anterior pituitary.

3. gonadotropin-releasing hormone (GnRH) – this hormone will travel to the anterior pituitary and cause more luteinizing hormone and follicle stimulating hormone to be released.

4. thyrotropin-releasing hormone (TRH) - this hormone travels to the anterior pituitary and will cause more thyroid stimulating hormone to be released.

5. corticotropin-releasing hormone (CRH) – this hormone travels to the anterior pituitary and will cause more adrenocorticotropic hormone to be released.

6. growth hormone-releasing hormone – (GHRH) – this hormone travels to the anterior pituitary and will cause more growth hormone to be released.

7. growth hormone-inhibiting hormone – (GHIH) (somatostatin) – this hormone travels to the anterior pituitary and will cause less growth hormone to be released.

Remember that these hormones all leave the hypothalamus, travel through the hypothalamohypophysial portal system, enter the anterior pituitary only, and then affect the cells of the anterior pituitary. These hormones have nothing to do with the posterior pituitary. Notice also in the names of these hormones you will see one of two words: releasing or inhibiting. So when you see releasing or inhibiting in the hormone name, think hypothalamus.

With all hormones you want to always know a few things

about them:

   1. production site – which means, "Where is the hormone made?"

   2. target tissue – which means, "What part of the body does the hormone affect?"

   3. effects – which means, "What will the hormone do, once it reaches the target tissue?"

With the seven hormones listed above, they all have a production site of hypothalamus, they have the anterior pituitary as the target tissue and the effects are different. These hormones are the only ones associated with the portal system and the name of the hormone tells you what its effects are. These are the simplest of the hormones.

POSTERIOR PITUITARY HORMONES

The hormones of the posterior pituitary are neurohormones. Remember that the posterior pituitary is just an extension of the hypothalamus. So the hormones released from the posterior pituitary are all made in the hypothalamus. Often you may be asked, "Where is oxytocin and ADH produced?" or "Where is oxytocin and ADH released?" With most hormones the answer is the same, but not with these two. The production site and the release site are different.

   1. oxytocin – oxytocin has two different target tissues. The first is the smooth muscle of the uterus. As a baby grows, it will stretch the smooth muscle of the uterus, which surrounds it. The stretching of the smooth muscle will cause the hypothalamus to release more oxytocin from the posterior pituitary. The more the baby grows, the more oxytocin is released. The oxytocin will leave the posterior pituitary, enter the blood and eventually reach the uterus. When it reaches the uterus, it will cause contractions. This chemical is the primary stimulus for the contractions seen at the time

of labor. After the baby is born, the uterus is no longer being stretched, so oxytocin release is slowed and the uterus stops contracting.

The other target tissue for oxytocin is the smooth muscle of the mammary glands. When a baby nurses, the hypothalamus will release more oxytocin from the posterior pituitary. This oxytocin will enter the blood and reach the mammary glands. The oxytocin will cause the smooth muscle of the mammary glands to contract and milk will be released. Sometimes when a baby is nursing, uterine contraction can be felt, because no matter the stimulus the hormone still affects the same tissues.

So oxytocin has two target tissues: uterus and mammary glands. Oxytocin also has two effects: contraction of the uterus and milk release.

2. antidiuretic hormone (ADH) – Antidiuretic hormone targets the kidneys and we will see where in particular, when we get to the kidney chapter. Think about what happens when someone takes a diuretic, they see increased urine output. You can also think of this as an increase in water loss. Why would the body want to lose water? To balance blood pressure. Always remember, water balance and blood pressure go hand and hand. When someone is holding a large volume of water, the blood pressure will be high. When a person has lost a large volume of water, the blood pressure will be low. ADH secretion is all about balancing blood pressure.

The water we see lost as urine is taken from the plasma of the blood, so our urine was plasma, just before it was produced. ADH causes water to be retained by increasing the reabsorption of sodium. Remember that water always follows solutes and sodium is the most abundant solute we have. Wherever sodium goes, water follows it. So, as ADH increases sodium reabsorption, we start holding the sodium. As the sodium is retained in the body, the water follows it and is reabsorbed back into the body. Why would the body want to reabsorb water? To increase blood pressure. When do you think the body would want to increase blood pressure? When blood pressure is low. Low blood pressure is the primary stimulus for the hormone ADH.

ADH is also called vasopressin because of a second target tissue, the smooth muscle around blood vessels. All blood vessels except for the smallest have smooth muscle around them. Think about where the word vasopressin comes from. Vaso for vessel and pressin for the pressing on the blood after causing smooth muscle to constrict around blood vessels. Remember that ADH causes the kidneys to hold sodium and water, and this will help to raise blood pressure. At the same time it will cause the smooth muscle around blood vessels to constrict, this will also raise blood pressure. So, as water is retained and blood vessels constrict, blood pressure rises. When would the body release ADH? When blood pressure is low.

Our body has two different mechanisms for regulating ADH secretion. One way is through the use of baroreceptors. From the nervous system you should have learned that baroreceptors are sensory receptors which detect pressure (or stretching whichever you prefer). Arteries are elastic, so when our blood pressure changes, they will stretch more or less. These baroreceptors can be found in places like the aorta and the internal carotid arteries. When our blood pressure drops, our arteries are being stretched only a small amount. When our blood pressure rises, our arteries are being stretched a large amount. This change in stretching, also seen as a change in pressure is being monitored by the hypothalamus. A small amount of stretching sends action potentials back to the brain infrequently, a large amount of stretching will send action potentials back to the brain more frequently. The brain sees a low frequency as low blood pressure and a high frequency as high blood pressure. So as baroreceptors detect that the arteries are being stretched only a small amount, this tells the brain that blood pressure is low. If blood pressure is low the hypothalamus will release ADH from the posterior pituitary. The ADH will cause the kidneys to hold water, raising the blood pressure and it will cause smooth muscles around blood vessels to constrict, raising blood pressure. This is simple negative feedback.

If blood pressure is high, the baroreceptors will detect this increase in pressure. This stimulus will slow the release of ADH. As less ADH is released, we stop holding water. As we stop holding water, we lower our blood pressure. At the same time smooth muscle around blood vessels aren't constricted. As the vessels enlarge, pressure drops. Think of less ADH as the same as taking a diuretic.

Other sensory receptors affecting ADH secretion are osmoreceptors. Osmoreceptors are found within neurons of the hypothalamus. To understand how these receptors work, we need to know what osmolality is. Osmolality means, "The number of particles in solution." They are basically viscosity receptors. When the hypothalamus detects a rise in osmolality, this means that our blood has too many solutes in it (it's too thick). This means our blood doesn't have enough water in it. If we don't have enough water in our blood, which means the blood has too many particles in solution, which means it is too thick, what do we want to do with water? We want to hold water. If we want to hold water, what do we do to ADH secretion? We release more ADH and this will cause our kidneys to hold water. If we are low on water, we want the kidneys to hold it. This is simple negative feedback.

At the same time our osmoreceptors are affecting ADH secretion, it will do something else. If again our blood doesn't have enough water in it, what do we want the kidneys to do with water? Hold it. As the kidneys are holding water, our hypothalamus will also tell us, we are thirsty. When we drink water, this will add water to the blood and thin it back out. Osmolality will be brought back into balance and homeostasis is maintained. Remember that the hypothalamus is where we have our thirst center located.

ANTERIOR PITUITARY HORMONES

1. Prolactin – Prolactin is another hormone involved with the mammary glands. Prolactin will stimulate the mammary glands to produce milk, not release it. So for the mammary glands to work properly prolactin is needed for milk production and oxytocin is needed for milk release.

2. Melanocyte stimulating hormone (MSH) – MSH targets the melanocytes of the skin. You may recall that melanocytes are the skin cells responsible for the production of melanin, which gives us our skin and hair color. The name of the hormone tells you exactly what it will do, cause melanocytes to produce more melanin. Usually our melanocytes are producing more melanin in response to exposure to ultraviolet light, but that is not the only time. During pregnancy this hormone level may rise and darken the face and chest of a pregnant woman.

3. Lipotropins – It doesn't take much imagination to guess that this chemical signal has something to do with lipids. The lipid storing cells of our body are called adipocytes. Lipotropins will target the adipocytes and tell them when to release fats. Fats are our number one energy storage molecule in the body. If we break down and release fats from the adipocytes, this is energy release. Our fat cells store energy for future use and lipotropins tell them when to release that stored energy. When would we want energy to be released? When our energy (blood sugar) levels are low. If we haven't eaten for several hours or maybe we have been working hard, our blood sugar levels drop. By releasing lipotropins, we can raise our blood sugar levels back up. Because hormones enter our blood and travel all over the body, this illustrates why it isn't possible to spot burn fats. We can't spot burn fats because lipotropins travel all through the body and affect adipocytes everywhere, not in just one place.

4. Beta endorphins – Beta endorphins are chemicals which work on the brain to stop pain. All sensations must take place in the brain, so if we want to stop a sensation, the brain is the place to do it. When would we see beta endorphins released? When we are in pain.

5. Growth Hormone – Growth hormone is a chemical signal which works on many tissues of the body, but not all. Growth hormone's name tells you just what it does, causes growth. This chemical signal is found in higher concentrations in our body, when we are young. For proper growth this is not the only hormone needed. In addition to growth hormone, we also need adequate levels of thyroid hormone, estrogen (females) and testosterone

(males). Growth hormone stimulates the uptake of amino acids for protein synthesis and will also cause the release of energy from adipocytes.

      6. Thyroid stimulating hormone (TSH) – TSH targets the cells of the thyroid gland and will cause the thyroid cells to release thyroid hormones (T3 and T4). Thyroid hormone will be discussed later.

      7. Adrenocorticotropic hormone (ACTH) – ACTH targets the cells of the adrenal cortex, which is part of the adrenal glands. The adrenal glands have two halves, much like the pituitary does. The cortex region will be affected by ACTH and will release hormones in response to it.

      8. Luteinizing hormone (LH)- LH targets the gonads of the body. The gonads are the organs of the body which produce the reproductive cells. The gonads include the ovaries of the female and the testes of the male. In females LH is the hormone in control of ovulation, which is the release of the female gamete called the oocyte. Also, in females it will cause development of the corpus luteum. In males LH will stimulate production of testosterone.

      9. Follicle stimulating hormone (FSH) – FSH also targets the gonads of the body. In females this chemical will cause the ovarian follicles to develop, which will at the same time cause the production of estrogen. In males this chemical will stimulate sperm cell production.

      When looking at the nine major hormones of the anterior pituitary, notice how they have many target tissues and effects.

      Now remember that the hormones of the anterior pituitary are produced by epithelial cells, not neurons, like the hormones of the posterior pituitary.

PINEAL GLAND

      The primary hormone of this gland is melatonin. Melatonin

secretion is regulated by photoperiod. Photoperiod refers to the number of daylight hours we have throughout the year. We all know that in the fall and winter the days get shorter. As the days get shorter, the pineal gland detects this through our vision. As the days get shorter, more melatonin is released. Melatonin works in a few major ways, one is to relax the individual. Think of melatonin as a chemical that mellows someone out, in other words, calms them. In the winter many animals are less active, because of the effects of melatonin. Animals need to be less active in the winter, because less food is available to them.

Another function of melatonin is to inhibit reproductive cycles. Again melatonin secretion is highest in the winter. Not many animals are going to reproduce in the winter. In the winter food is scarce and newborns might freeze to death.

As we approach the spring and summer, the days get longer. At this time less melatonin will be released and animals will become more active, because melatonin is not there to slow them. Also consider when do most animals reproduce? In the spring and summer. Why? This is when melatonin levels are low. If melatonin is not there to inhibit reproduction, then reproduction occurs. This is the part of the brain that is often referred to as the biological clock.

THYROID GLAND

The thyroid gland is located just inferior to the larynx (voice box) and anterior to the trachea. It's shaped like a bow tie and has a narrow connection in the center called the isthmus.

When looking at a histology slide of the thyroid gland, you will see two things. There will be a number of large follicles, which look like hollow spheres. These follicles store up the hormones T3 and T4. The storage of a hormone is rare for endocrine glands. In between these large follicles parafollicular cells will be seen. From these parafollicular cells comes the hormone calcitonin.

1. Thyroid hormone (T3 and T4) – Thyroid hormone has the

very important job of regulating metabolism. Metabolism can be defined as the sum of all chemical reactions occurring inside the body. Largely what these hormones will do is to increase the production of mitochondria in cells and also makes the mitochondria more active. If cells have more mitochondria and they become more active, then the cells will have much more ATP available to them. With more ATP the cells will be more active, thus increased metabolism.

The symbol T3 stands for triiodothyronine and T4 stands for tetraiodothyronine (thyroxine). The 3 and the 4 refer to how many iodine atoms each molecule contains. We must have iodine in our diet if we wish to produce these hormones. Iodine is placed into salt to prevent deficiencies of this hormone. If a mother doesn't have adequate amounts of iodine in her diet during pregnancy, the child may be born with cretinism. This is a form of mental retardation, where the nervous system doesn't develop properly, due to a lack of thyroid hormone. The nervous system develops faster than the other systems of the body. Thyroid hormone is needed for tissues to mature, so without it, the nervous system doesn't develop properly.

Thyroid hormone targets most tissues of the body and is needed for proper growth and development.

2. Calcitonin – Anyone could quickly figure out that calcitonin has something to do with calcium. This is one of the major regulators of blood calcium levels.

Calcium is needed for much more than strong bones. It is needed for membrane potentials, muscle contraction, blood clotting, etc.

Calcitonin targets the osteoclasts of our bones. Any cell with clast on the end of it will be a cell that breaks down a tissue, so osteoclasts break down bone. Remember that bone is a connective tissue and one of the functions of connective tissue is the storage of materials. Most everything in a tissue came from the blood, so if a tissue is broken down, it will send its materials back into the blood.

Bone stores calcium for later use. If our blood levels become low, bone can release stored calcium. Calcitonin will inhibit osteoclast activity. Think of this chemical as an off switch for osteoclasts. Since osteoclasts break down bone, which causes a release of calcium, calcitonin will stop that release of calcium. When would we want to stop osteoclasts and stop releasing calcium? When our blood calcium levels are adequate or high. So calcitonin works to lower blood calcium levels.

PARATHYROID GLANDS

The parathyroid glands are located on the posterior surface of the thyroid gland. They will be at the corners of the bow tie shaped thyroid gland.

The parathyroid glands produce a hormone called parathyroid hormone (PTH). PTH works just the opposite of calcitonin and also has more target tissues. PTH will increase osteoclast activity. If osteoclast activity is increased, then more calcium will be released from the bone and into the blood. When would we want to release more calcium into the blood? When blood calcium levels are low. So PTH works to raise blood calcium levels.

In addition to the osteoclasts, PTH also targets the small intestine. PTH will cause the cells of the small intestine to absorb calcium, we have consumed in our diet. This will also elevate blood calcium levels.

PTH will also target the kidneys. PTH will cause the kidneys to reabsorb more calcium and prevent calcium loss in the urine. This will also help to elevate blood calcium levels.

PTH will also increase the production of vitamin D. Vitamin D will increase the absorption of calcium in the small intestine, which will also work to raise blood calcium levels.

Calcitonin and PTH work together to balance our blood calcium levels. When one of these hormones is increasing, the other will be decreasing.

ADRENAL GLANDS

The adrenal glands are found on the superior poles of the kidneys. Since we have two kidneys, we have two adrenal glands. The adrenal glands have two halves, which are very similar to the halves of the pituitary gland. One half is the adrenal cortex, which is composed of epithelial cells. The other half is the adrenal medulla, which is part of the sympathetic nervous system.

Adrenal medulla - Since the adrenal medulla is part of the sympathetic nervous system, then it must release epinephrine and norepinephrine. You should have heard of the sympathetic division in the nervous system chapter. Epinephrine and norepinephrine target many tissues of the body. Some tissues and organs will be stimulated and others will be inhibited. Remember that the sympathetic division stimulates organs needed for physical activity and inhibits many structures not needed for physical activity. Epinephrine and norepinephrine will increase heart rate, increase our respiratory rate, increase blood flow to skeletal muscles and have many other effects. At the same time they will inhibit the digestive system, inhibit the reproductive system, decrease blood flow to these organs and more.

Adrenal cortex – The adrenal cortex will release different hormones than the medulla and is not part of the nervous system in any way.

1. aldosterone (mineralocorticoids) – Aldosterone targets the kidneys and has several effects. One effect is to increase sodium reabsorption in the kidneys. If more sodium is reabsorbed, then more water will be absorbed. This effect will help to decrease urine output, which will work to increase blood pressure.

Aldosterone will also increase the secretion of hydrogen and potassium ion. By balancing hydrogen ions the body can maintain pH balance. The body has many ways to balance the pH, this hormone is just one. If a person develops acidosis, then they would want to release aldosterone, which would cause hydrogen ions to be lost in the urine. This would help to bring pH balance back. If a person has too much potassium in the blood they would want to release more aldosterone and this would remove the excess potassium with the urine.

2. cortisol (glucocorticoids) – Usually when you hear of this hormone you will hear about its anti-inflammatory effects. If a person has chronic inflammation in a joint, this chemical is often used for treatment.

3. androgens – Androgens help to support the functions of the female reproductive system.

PANCREAS

The pancreas is found in the upper left quadrant of the abdomen and is posterior to the stomach. It is also a mixed gland, meaning a gland that is part endocrine and part exocrine. The anatomy is simple; it has a head, body and tail. The head lies in the bend of the duodenum and most of the body and tail are posterior to the stomach. Within the endocrine part of the pancreas are three major cell types. The alpha cells secrete glucagon, the beta cells secrete insulin and the delta cells secrete somatostatin. The pancreatic islets are collections of endocrine cells found within the pancreas.

The major hormones of the pancreas are:

1. Insulin – Insulin targets many cells of the body and will promote the intake of glucose, and amino acids. When we eat a meal our blood sugar levels rise. The beta cells will detect the rise in blood sugar and in response, insulin will be released. Most cells of the body will respond to this insulin by taking in blood sugar and amino acids. As blood sugar levels start to drop, insulin secretion

will slow. So insulin works to lower our blood sugar levels.

We mentioned that insulin doesn't affect all cells of the body. Neurons don't wait on insulin to tell them to take in blood sugar. Neurons are the most active cells of the body, so they must always take in sugar. Consider what would happen if a diabetic injects too much insulin. If too much insulin is taken, then cells will absorb too much sugar. If this happens there may not be enough sugar in the blood to keep the neurons supplied with energy. If the neurons run out of energy the nervous system can fail and the person could die.

The liver and adipocytes are very receptive to insulin. The liver and adipocytes are energy storage sites in the body. When would we want these organs to store up energy? After we eat a meal when blood sugar levels are high is the best time. So when we eat and have a large amount of sugar in the blood, the liver and adipocytes will store up energy. The energy will be stored in the form of glycogen in the liver and fat in the adipocytes.

Another part of the body that responds to insulin is the hypothalamus. Within the hypothalamus is the satiety center and the hunger center. The role of insulin on these parts of the brain is not clearly understood.

2. Glucagon – Glucagon is secreted by the alpha cells of the pancreas. Glucagon will target the energy storage sites of the body and signal them to release energy. As the energy storage sites release energy, blood sugar levels will rise. When would we want to raise blood sugar levels? When our blood sugar levels are low. Glucagon will be released when we haven't eaten for some time and our blood sugar levels have dropped. The liver and adipocytes stored energy in response to insulin and they will release it in response to glucagon. Notice glucagon doesn't have as many target tissues as insulin, because not many cell types have the function of energy storage and release.

Notice how insulin and glucagon have antagonistic effects on each other to maintain blood sugar level. When one of these hormones is being stimulated the other will be inhibited.

3. Somatostatin – Somatostatin is produced by the delta cells of the pancreas. Somatostatin inhibits insulin and glucagon secretion. Some of its roles appear unclear.

Whenever a discussion of insulin and blood sugar comes up the topic of diabetes will arise. You should have an understanding of diabetes. Diabetes comes in two major forms: diabetes insipidus and diabetes mellitus.

Diabetes insipidus involves a problem with the hormone ADH. Diabetes insipidus comes in four forms: neurogenic, nephrogenic, gestagenic and dipsogenic. You are unlikely to encounter anyone with this form of diabetes.

What you have almost always heard of is diabetes mellitus. Diabetes mellitus comes in two major forms: type I and type II.

Type I diabetes mellitus is also called juvenile diabetes, because it often develops early in life. This form of diabetes is a bit rare and accounts for 3% of all diabetes cases. This diabetes is caused by an autoimmune disease, meaning that for some reason the immune system has destroyed the insulin producing cells of the pancreas. If these cells are destroyed then the body no longer produces insulin. Without insulin many cells of the body won't take in blood sugar.

Type II diabetes mellitus is also called adult onset diabetes, because it often develops later in life, often in the forties and above. This diabetes accounts for about 97% of all diabetes cases. Lifestyle is the reason this diabetes is so common. Many people choose a poor diet and lack of exercise through most of their life. If a person is taking in a large amount of calories and not exercising, what will happen to insulin levels? If blood sugar is always high, then insulin levels will always be high. Over years the cells of the body will become tolerant to the insulin, in other words, they stop responding to it. If cells stop responding to insulin, they stop taking in blood sugar. Over years a poor diet can cause a clogging of the arteries and at the same time chronic high blood sugar will cause a loss of circulation. When circulation is inhibited, all organs will be affected. Many people have amputations, kidney failure, blindness and many other problems.

THYMUS GLAND

The thymus gland is located just superior and anterior to the heart. This gland plays a large role in the lymphatic system, but it does release one or more hormones associated with the development of the lymphatic system. The hormone thymosin is believed to be involved with the maturing of T-cell lymphocytes.

TESTES

Testes are the paired gonads of the male reproductive system. Within the testes are a collection of cells called the interstitial or (Leydig) cells. The Leydig cells produce the hormone testosterone for the male. Testosterone affects many tissues of the body. Testosterone causes enlargement of the larynx, increased muscle mass, development of the male reproductive system and many more effects. This hormone is also needed for production of the male reproductive cells, sperm cells.

OVARIES

The ovaries are the paired gonads of the female reproductive system. The major hormones of the ovaries are estrogen and progesterone. Estrogen causes a maturing of the female reproductive system and menstrual cycle. Progesterone is also involved with the menstrual cycle and the endometrium.

We have been over the major endocrine organs of the body, but there are many other very important hormones in the body. Organs that you may not think of as having endocrine functions do. A few of these organs are the heart and kidneys. These organs are generally not considered part of the endocrine system, but they do release important hormones.

HEART

The heart produces a hormone called atrial natriuretic hormone (ANH). ANH is released from the right atrium when the walls of the atria are being stretched under high blood pressure. ANH targets the kidneys and decreases the reabsorption of sodium. When more sodium is lost, so is water. This hormone will increase urine output and as urine output increases, blood volume decreases. This hormone is released at times of high blood pressure and as water is lost, blood pressure drops. This hormone has the opposite effects of ADH.

KIDNEYS

One of the major hormones of the kidneys is erythropoietin. The kidneys monitor blood oxygen levels and can detect when oxygen delivery is inadequate. If oxygen delivery is inadequate the hormone erythropoietin is released. This hormone targets the red bone marrow and will increase red blood cell production. An increase in red blood cells will result in increased oxygen delivery. This would be an example of negative feedback.

The digestive system releases many hormones important to the digestive system. Some hormones are seen only at the time of pregnancy and the list goes on.

Use the following table as a study aid. If you can memorize the following table, you will know much of the important hormone information.

Hormone	Production Site	Target Tissue	Response
PRH	hypothalamus	anterior pituitary	increased prolactin secretion
PIH	hypothalamus	anterior pituitary	decreased prolactin secretion
GnRH	hypothalamus	anterior pituitary	incr. secretion of LH & FSH
TRH	hypothalamus	anterior pituitary	incr. TSH secretion
CRH	hypothalamus	anterior pituitary	incr. ACTH hormone secretion
GHRH	hypothalamus	anterior pituitary	incr. growth hormone secretion
GHIH	hypothalamus	anterior pituitary	decr. growth hormone secretion
Oxytocin	hypothalamus	mammary glands, uterus	milk release uterine contractions
ADH	hypothalamus	kidneys	incr. water reabsorption
Prolactin	anterior pituitary	mammary glands	milk production
MSH	anterior pituitary	melanocytes	incr melanin production
Lipotropins	anterior pituitary	adipocytes	incr. fat breakdown

Hormone	Production Site	Target Tissue	Response
B endor	anterior pituitary	brain	pain killer
GH	anterior pituitary	most tissues	growth, amino acid intake
TSH	anterior pituitary	thyroid gland	incr thyroid hormone
ACTH	anterior pituitary	adrenal cortex	incr glucocorticoids
LH	anterior pituitary	ovaries, testes	ovulation, testosterone synthesis
FSH	anterior pituitary	female follicles, seminiferous tubules	follicle maturation, and estrogen secretion in females sperm production
melatonin	pineal gland	hypothalamus	inhibits reproduction, affects sleep wake cycles
thyroid hormone T3, T4	thyroid gland	most cells	increased metabolism growth, maturation
Calcitonin	thyroid gland	bone	decr. osteoclast activity

Hormone	Production Site	Target Tissue	Response
PTH	parathyroids	bone, kidney small intestine	incr. osteoclast activity calcium absorption in kidneys and small intestine, Vit D prod.
Epinephrine Norepinephrine	adrenal medulla	many tissues	many effects
Aldosterone	adrenal cortex	kidneys	sodium reabsorption, hydrogen & potassium excretion
cortisol	adrenal cortex	many tissues	anti-inflammatory
androgens	adrenal cortex	many tissues	sex characteristics in females
insulin	pancreas	liver, adipocytes skeletal muscle	glucose and amino acid intake
glucagon	pancreas	liver, adipocytes	energy release
somatostatin	pancreas	alpha & beta cells	inhibits insulin & glucagon

Hormone	Production Site	Target Tissue	Response
thymosin	thymus gland	lymphatic system	matures lymphatic system
Testosterone	testes	many tissues	development of reproductive system, spermatogenesis
estrogen	ovaries	many tissues	development of reproductive system, menstrual cycle
progesterone	ovaries	many tissues	development of reproductive system, menstrual cycle
ANH	heart	kidneys	water loss
Erythropoietin	kidneys	red bone marrow	incr. red blood cell production

Chapter 1 – Study Questions

1. Hormones work on specific areas called?
a. ligands
b. pheromones
c. target tissues
d. baroreceptors
e. portals

2. What type of chemical signal acts locally and on the same cell type which secreted it?
a. autocrine
b. paracrine
c. hormone
d. neurotransmitter
e. pheromone

3. What type of chemical signal acts locally and on a different cell type than the secreting cell?
a. autocrine
b. paracrine
c. hormone
d. neurotransmitter
e. pheromone

4. What type of chemical signal enters the blood and travels some distance to its target tissue?
a. autocrine
b. paracrine
c. hormone
d. neurotransmitter
e. pheromone

5. What type of chemical signal is released into the synapse between neurons?
a. autocrine
b. paracrine
c. hormone

d. neurotransmitter
e. pheromone

6. What type of chemical signal is secreted into the environment and affects other individuals?
a. autocrine
b. paracrine
c. hormone
d. neurotransmitter
e. pheromone

7. What two body systems are the controlling systems of the body?
a. endocrine and cardiovascular
b. cardiovascular and nervous
c. lymphatic and endocrine
d. endocrine and nervous
e. none of the above

8. What type of chemical signal will pass freely through the plasma membrane?
a. lipid soluble
b. water soluble
c. both
d. neither

9. A water soluble chemical signal is most likely to work were?
a. the inside of the cell
b. the outside of the cell
c. both
d. neither

10. A lipid soluble chemical signal is most likely to work were?
a. the inside of the cell
b. the outside of the cell
c. both
d. neither

11. Most hormones are regulated by?
a. positive feedback

b. negative feedback
c. opposite feedback
d. none of the above

12. Which hormone is an example of the nervous system regulating the release of a hormone?
a. insulin
b. thyroid hormone
c. norepinephrine
d. calcitonin
e. testosterone

13. When one hormone regulates the release of another hormone, this is called a?
a. lipid hormone
b. water hormone
c. tropic hormone
d. plasma hormone
e. neurohormone

14. What hormone has a very rapid effect and a short half life?
a. thyroid hormone
b. estrogen
c. epinephrine

15. What hormone follows very regular patterns?
a. thyroid hormone
b. estrogen
c. epinephrine

16. What hormone follows a very constant patters and changes little over time?
a. thyroid hormone
b. estrogen
c. epinephrine

17. The hypothalamus connects to the anterior pituitary by the?
a. hypothalamohypophysial portal system

b. hypothalamohypophysial tract system
c. through both
d. neither

18. The hypothalamus connects to the posterior pituitary by the?
a. hypothalamohypophysial portal system
b. hypothalamohypophysial tract system
c. through both
d. neither

19. A portal system is?
a. a bundle of axons
b. two capillaries with a direct connection
c. a combination of the two
d. neither

20. The hypothalamohypophysial tract system is?
a. a bundle of axons
b. two capillaries with a direct connection
c. a combination of the two
d. neither

21. What is also known as the neurohypophysis?
a. anterior pituitary
b. infundibulum
c. hypothalamus
d. posterior pituitary
e. none of the above

22. The hormones which travel from the hypothalamus to the anterior pituitary will always have what in the name?
a. stimulating or inhibiting
b. stimulating or releasing
c. releasing or inhibiting
d. portal
e. tract

23. The posterior pituitary produces how many hormones?

a. 0
b. 1
c. 2
d. 7
e. 9

24. The posterior pituitary releases how many hormones?
a. 0
b. 1
c. 2
d. 7
e. 9

25. The anterior pituitary produces and releases how many hormones?
a. 0
b. 1
c. 2
d. 7
e. 9

26. Which of the following is a neurohormone?
a. ADH
b. MSH
c. prolactin
d. growth hormone
e. thyroid stimulating hormone

27. Which hormone can cause uterine contractions and milk release?
a. prolactin
b. follicle stimulating hormone
c. ACTH
d. luteinizing hormone
e. oxytocin

28. A nursing baby is likely to cause the release of?
a. prolactin

b. follicle stimulating hormone
c. ACTH
d. luteinizing hormone
e. oxytocin

29. A baby growing inside the uterus will cause the release of?
a. prolactin
b. follicle stimulating hormone
c. ACTH
d. luteinizing hormone
e. oxytocin

30. The release of ADH will cause urine output to?
a. decrease
b. increase
c. no effect

31. If the body releases less ADH, urine output will?
a. decrease
b. increase
c. no effect

32. If a person is losing a large amount of blood, you would expect ADH secretion to?
a. decrease
b. increase
c. no effect

33. If a person has chronic high blood pressure, you would expect ADH secretion to?
a. decrease
b. increase
c. no effect

34. If a physician wishes to prevent a woman from going into labor, the physician would need to block what hormone?
a. prolactin
b. follicle stimulating hormone

c. ACTH
d. luteinizing hormone
e. oxytocin

35. ADH will do what to sodium and water absorption in the kidneys?
a. decrease
b. increase
c. no effect

36. ADH attempts to do what to blood pressure?
a. decrease
b. increase
c. no effect

37. Which hormone will increase milk production?
a. oxytocin
b. prolactin
c. ADH
d. luteinizing hormone
e. lipotropins

38. Which hormone will darken the skin?
a. lipotropins
b. MSH
c. beta endorphins
d. thyroid stimulating hormone
e. follicle stimulating hormone

39. Which hormone would be released if a person is starving?
a. lipotropins
b. MSH
c. beta endorphins
d. thyroid stimulating hormone
e. follicle stimulating hormone

40. If a person has fractured a humerus, you would expect what hormone to be released?

a. lipotropins
b. MSH
c. beta endorphins
d. thyroid stimulating hormone
e. follicle stimulating hormone

41. Which hormone is responsible for raising the metabolism?
a. lipotropins
b. MSH
c. beta endorphins
d. thyroid hormone
e. follicle stimulating hormone

42. Which hormone will increase the number of mitochondria within our cells and make the mitochondria more active?
a. lipotropins
b. MSH
c. beta endorphins
d. thyroid hormone
e. follicle stimulating hormone

43. If a person is exercising intensely every day you would expect what hormone to be elevated?
a. lipotropins
b. MSH
c. beta endorphins
d. thyroid hormone
e. follicle stimulating hormone

44. What hormone targets the adrenal cortex?
a. LH
b. TSH
c. ACTH
d. GH
e. FSH

45. When a person uses a test kit to determine the best time to become pregnant, the kit will check for elevated levels of what hormone?
a. LH
b. TSH
c. ACTH
d. GH
e. FSH

46. The pineal gland releases which hormone?
a. calcitonin
b. T3 and T4
c. PTH
d. melatonin
e. glucagon

47. Which hormone requires iodine in its production?
a. calcitonin
b. T3 and T4
c. PTH
d. melatonin
e. glucagon

48. Which hormone works to decrease blood calcium levels?
a. calcitonin
b. T3 and T4
c. PTH
d. melatonin
e. glucagon

49. Which hormone will inhibit osteoclast activity?
a. calcitonin
b. T3 and T4
c. PTH
d. melatonin
e. glucagon

50. Which hormone works to increase blood calcium levels?

a. calcitonin
b. T3 and T4
c. PTH
d. melatonin
e. glucagon

51. Which of the following is not a target tissue of PTH?
a. small intestine
b. kidneys
c. osteoclasts
d. blood
e. none of the above

52. Which organ will be found on the superior poles of the kidneys?
a. adrenals
b. pancreas
c. thymus
d. pituitary
e. none of the above

53. The adrenal medulla is part of the?
a. central nervous system
b. peripheral nervous system
c. parasympathetic nervous system
d. sympathetic nervous system
e. all of the above

54. The adrenal medulla will release?
a. cortisol
b. aldosterone
c. epinephrine
d. androgens
e. all of the above

55. Activation of the adrenal medulla is likely to cause?
a. stimulation of the GI tract
b. urination
c. low blood pressure

d. sleep
e. rapid heart rate

56. If a person has acidosis, you would expect aldosterone secretion to?
a. decrease
b. increase
c. no effect

57. If a person was low on potassium, you would expect aldosterone secretion to?
a. decrease
b. increase
c. no effect

58. Which hormone would be used to treat inflammation?
a. aldosterone
b. cortisol
c. androgens
d. all of the above
e. none of the above

59. After eating a meal, you would expect insulin secretion to?
a. decrease
b. increase
c. no effect

60. Hours after a meal you would expect insulin levels to be?
a. high
b. low

61. Insulin works to do what to blood sugar levels?
a. raise
b. lower
c. no effect

62. Glucagon works to do what to blood sugar levels?
a. raise

b. lower
c. no effect

63. After hours of intense exercise you would expect glucagon levels to be?
a. high
b. low

64. Which would be a target tissue for glucagon?
a. heart
b. lungs
c. kidneys
d. liver
e. brain

65. The most common form of diabetes is?
a. diabetes insipidus
b. type 1 diabetes mellitus
c. type 2 diabetes mellitus

66. Type 1 diabetes mellitus is caused by?
a. poor diet and lack of exercise
b. an autoimmune disease
c. a problem with ADH

67. A person is no longer producing insulin with what form of diabetes?
a. diabetes insipidus
b. type 1 diabetes mellitus
c. type 2 diabetes mellitus

68. Thymosin is needed for proper functioning of what system?
a. cardiovascular
b. nervous
c. lymphatic
d. urinary
e. reproductive

69. Atrial natriuretic hormone will do what to urine output?
a. raise
b. lower
d. no effect

70. Atrial natriuretic hormone is released when blood pressure is?
a. high
b. low

71. Erythropoietin is released when oxygen delivery is?
a. high
b. low

72. The target tissue for erythropoietin is?
a. ovaries
b. testes
c. kidneys
d. nervous system
e. bone marrow

Answers to multiple choice questions.
1. C
2. A
3. B
4. C
5. D
6. E
7. D
8. A
9. B
10. A
11. B
12. C
13. C
14. C
15. B
16. A
17. A
18. B
19. B
20. A
21. D
22. C
23. A
24. C
25. D
26. A
27. E
28. E
29. E
30. A
31. B
32. B
33. A
34. E
35. B
36. B
37. B

38. B
39. A
40. C
41. D
42. D
43. D
44. C
45. A
46. D
47. B
48. A
49. A
50. C
51. D
52. A
53. D
54. C
55. E
56. B
57. A
58. B
59. B
60. B
61. B
62. A
63. A
64. D
65. C
66. B
67. B
68. C
69. A
70. A
71. B
72. E

BLOOD

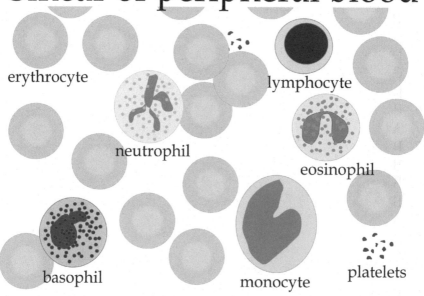

Smear of peripheral blood

CHAPTER 2
BLOOD

The blood is a type of connective tissue and it connects almost all tissues of the body. Some tissues are avascular, but most do contain blood vessels. The blood is what brings in nutrients, gases, hormones, ions, proteins and other materials. Equally important are the materials the blood removes. Wastes must be removed if cells are to live.

The blood is also responsible for moving heat around the body. As our blood travels deep, it picks up heat. As it travels superficially, where things are cooler, the blood will release heat. The blood vessels in the skin act like the radiator on a car. When a car engine heats up, it will begin to pump blood through the radiator, so heat will be released. As we begin to heat up, the blood vessels in our skin will dilate. The dilation brings more blood to the surface of the body, which releases more heat. This dilation can be seen, when a person's face is red.

Blood Composition

Blood can be broken down into two materials: formed elements (cells) and plasma (water and solutes). These two materials can be broken down further. Make sure you know the composition of blood.

Formed elements – The formed elements are the cells of the blood. The blood cells are broken down into three main types: erythrocytes, leukocytes and thrombocytes.

ERYTHROCYTES

Red blood cells (erythrocytes) – Red blood cells make up about 45% of the blood. The percent that red blood cells makeup of the total blood volume is called the hematocrit. It is vital to maintain a normal hematocrit if homeostasis is to be maintained. If hematocrit is too high, then that means the blood contains too many red blood cells. If the blood has too many red blood cells, then it will be too thick. The problem with thick blood is that the heart has to pump it. The thicker the blood, the more energy the heart has to expend to move it. We also don't want to have too few red blood cells. If we don't have enough red blood cells, oxygen delivery will be inadequate. Red blood cells are the most abundant of the formed elements and males will have around 5.5 million per cubic millimeter and women will have 4.8 million.

The size of our red blood cells is also important to us. If red blood cells are too small, then they won't be transporting enough oxygen. If our red blood cells are too large, they won't fit through the tiny capillaries. Normal red blood cells should be 7.5 microns in diameter. Capillaries are 7-10 microns in diameter. The capillary can clog up and oxygen delivery will stop, if the red blood cells don't bend and flex through them.

Red blood cells must also have a normal round, biconcave shape. Biconcave means they are thin in the center and thick around the edges. This shape will allow them to bend and flex easily when they go through small spaces. This shape not only makes them flexible, it also increases surface area. More surface area means more gas exchange. Red blood cell shape becomes a problem in people with sickle cell anemia. With sickle cell anemia the red blood cells take on a crescent shape. This shape is an adaptation to many generations being exposed to the malaria parasite. This parasite must have a normal round red blood cell to infect, so people exposed to the parasite have adapted. The new crescent shape helps to protect them from the parasite, but the crescent shape doesn't bend and flex the way a round cell will. Because of this the red blood cells won't fit through tiny spaces and circulation will sometimes stop in a tissue.

Our red blood cells transport almost all of the oxygen, moved around the body. We need this oxygen for aerobic respiration, which is the process by which we get most all ATP. The red blood cells also move about a quarter of the carbon dioxide around the body. We must remove carbon dioxide from tissues or the tissue will quickly become acidic. When carbon dioxide is produced, it will mix with water to make carbonic acid. This carbonic acid will dissociate to make hydrogen ion and bicarbonate ion. The buildup of the hydrogen ion is what will cause acidosis.

In the mature form red blood cells don't have a nucleus. The nucleus is lost to make more room for hemoglobin. Hemoglobin is the oxygen transporting molecule in the red blood cell and it is what makes the cell appear red. To make a functioning hemoglobin molecule, iron is needed and iron is what will transport the oxygen.

Erythropoiesis is the production of red blood cells and this takes place in the red bone marrow. The kidneys have the job of monitoring red blood cell production. If the kidneys determine that oxygen delivery is inadequate, they will release the hormone erythropoietin. This hormone will be transported to the red bone marrow and cause stem cells to develop into red blood cells. This production of red blood cells is called erythropoiesis. We must produce a few million red blood cells every second, just to replace the number lost naturally.

Anemia can be defined in two ways. Anemia can be a lack of red blood cells or a lack of hemoglobin. Either way oxygen delivery will be inadequate. Pernicious anemia is caused by a lack of B12, folate anemia is a lack of folic acid and aplastic anemia is often caused by chemicals or some other substance destroying the red bone marrow.

Polycythemia is an excess of RBC. High RBC counts are often due to inadequate oxygen delivery for some reason. When we have too many RBC our blood becomes too thick and the heart has to work harder to pump it.

Red blood cell facts to remember.

1. Red blood cells make up 45% of the blood on average.

2. Hematocrit – The percent of total blood volume that red blood cells make. Normal hematocrit is about 45%.

3. Red blood cell numbers should be: males 5.5 million per cubic millimeter, females 4.8 million per cubic millimeter.

4. Normal size for a red blood cell is 7.5 microns.

5. RBC should have a biconcave shape, meaning thin in the center and thick around the edges.

6. RBC shape is important for two reasons. It makes the cell flexible and increases surface area. Flexibility is needed to bend through tight spaces. Increased surface area allows for more oxygen exchange.

7. Primary function of RBC is to transport oxygen. They also transport about ¼ of the carbon dioxide.

8. RBC have a nucleus early in development, but don't have a nucleus in the adult form, we see in the blood. The nucleus is lost to make room for more hemoglobin. More hemoglobin means more oxygen transport and delivery.

9. RBC must have iron to make a hemoglobin molecule.

10. Erythropoiesis is the production of red blood cells.

11. Erythropoietin is the hormone produced by the kidneys which causes the red bone marrow to make more RBC.

12. Anemia is defined as a lack of RBC or a lack of hemoglobin.

13. RBC live about 110-120 days. Most are lost in the spleen, liver and capillaries. Macrophages in the spleen and liver will recycle and clean up the lost material. These waste materials will be removed by the liver and kidneys. The lost pigments like bilirubin give urine its yellow color.

14. We lose about .87% of our RBC each day. That is less than 1% per day.

15. Jaundice is the accumulation of bile pigments in the skin. This can be caused by liver damage, blockage of the bile duct or the

destruction of a large number of red blood cells. The yellow color of urine and the brown color of feces comes from the waste products of destroyed RBC.

16. Polycythemia is an excess of RBC.

LEUKOCYTES

White blood cells (leukocytes) – We have five major types of white blood cells in the body. These five are separated into two groups, the granulocytes and agranulocytes. The granulocytes get their name because of the granules which may be seen inside of them. The granulocytes are neutrophils, eosinophils and basophils. The agranulocytes are lymphocytes and monocytes.

Unlike red blood cells, which lose their nucleus as they mature, white blood cells retain the nucleus throughout life. All leukocytes have one nucleus but they will sometimes appear to have more than one. The reason is that the nucleus of the granulocytes has more than one lobe. The lobes will often appear to be unconnected when looking at them with a light microscope, but the connections are there. The shape and number of the lobes is one way white blood cells can be distinguished.

Where red blood cells can't move on their own, white blood cells can. Many leukocytes won't stay in the blood for long periods of time. They will often migrate to other tissues, where they can remove and destroy; just about anything we don't want in our bodies. Diapedesis is the process by which white blood cells leave the blood and enter the tissues. Don't confuse this process with chemotaxis. Chemotaxis is the process by which white blood cells follow chemical trails to areas of damage. Foreign invaders and inflammation will cause the release of chemicals. White blood cells will follow these trails back to the source, where the WBC are needed. Think of a dog following a scent trail to a deer. The white blood cells are hunting foreign invaders, so they will follow the chemical trails back to their source in pursuit of the invader.

White blood cells are normally found in numbers between 5000 and 9000 per cubic millimeter. The numbers of white blood cells does have clinical significance, but the percent that each WBC makes of the total WBC number is more significant. Make sure you know the normal percent for each WBC and what is the significance of any one being elevated.

Neutrophils are the most abundant of the white blood cells. Neutrophils will usually comprise 60-70% of all white blood cells. They are often easily identified, because of their polymorphonuclear shape. This means that they have many shapes to their nucleus. The nucleus of the neutrophil will often have 3-5 lobes (sometimes 2) in the nucleus. This separates them from the other white blood cells and often makes them easily identifiable. Neutrophils will quickly leave the blood and enter the tissues. They will search out anything not belonging in the body and phagocytize it (eat it). These WBC will only live for a day or so and are usually the first seen at sights of infection. At sights of infection we often see pus. Pus is an accumulation of dead WBC and most of the pus will be dead neutrophils. Neutrophils are associated with acute bacterial infections and appendicitis. So if high numbers of this WBC are seen in the blood, then a person may have an acute bacterial infection or appendicitis.

Basophils are another granulocyte found in the blood. Basophils are very rare and only account for .5-1% of all WBC. They are identified by the presence of two dissimilar lobes in the nucleus. So look for two lobes, which aren't identical. They will sometimes have a nucleus shaped like the letter S. Basophils are capable of releasing two very important chemicals, heparin and histamine. Heparin is a powerful anticoagulant. Anticoagulants will prevent clot formation where it isn't needed. A clot forms when certain chemicals are released in large numbers. The heparin will prevent these chemicals from spreading far from the production site, thus preventing clots from forming in places we don't need them. Histamine promotes inflammation and we need inflammation when tissues are damaged. Inflammatory chemicals largely work to increase blood flow to an area and attract WBC. Allergic reactions are associated with basophils, because of the inflammation. Basophils are commonly high in number after radiation exposure.

Eosinophils are the third granulocyte found in the blood. Eosinophils are relatively rare and account for 2-4% of all WBC. They are identified by the presence of two identical lobes in the nucleus. Often the nucleus will look like two eyes, looking back at you. In addition the eosinophils will often appear very grainy, while the other granulocytes don't. Eosinophils release chemical which prevent inflammatory chemicals from spreading, thus reducing inflammation. So when a large number of these cells are seen, they are usually reducing inflammation. In addition, eosinophils produce chemicals used to destroy parasites.

Lymphocytes are an agranulocyte, meaning it doesn't have the granules seen in other WBC. Lymphocytes are common in the blood and make 20-25% of all WBC. The lymphocytes are the smallest of the WBC and don't have lobes in their nucleus. The nucleus will appear as one large round structure, which occupies most of the cell. Lymphocytes will leave the blood and migrate to the lymphatic tissues, which is where most of them are found. There are many types of lymphocytes, but the two main groups are B cells and T cells. The B cells are responsible for antibody production and the T cells work to destroy viruses and cancers. Much more will be covered on these cells in the lymphatic system.

Monocytes are the second agranulocyte and are the largest of the WBC. Monocytes make 3-8% of all WBC and will quickly leave the blood. They are identified by their large size and will sometimes have a horseshoe shaped nucleus. After monocytes leave the blood and enter a tissue, they will enlarge and become a macrophage. The macrophage will wander the tissues and remove anything which doesn't belong. Where some cells like neutrophils die after one phagocytic event, macrophages don't. They will live much longer, so they are associated with chronic infections.

Too few or too many WBC can indicate a disease condition within the body. Leukopenia is an unusually low number of WBC.

Leukocytosis is an unusually high number of WBC.

WBC facts to remember.

1. We have five major types of WBC in the body.

2. All WBC fall into two groups, granulocytes and agranulocytes.

3. WBC always retain their nucleus, unlike RBC.

4. WBC can move on their own by ameboid movement, where RBC can't.

5. Diapedesis is the process by which white blood cells leave the blood and enter the tissues.

6. Chemotaxis is the process by which white blood cells follow chemical trails to areas of damage.

7. Neutrophils are the most abundant of WBC, make 60-70% of all WBC, usually have 3-5 lobes, and will be associated with acute bacterial infections and appendicitis.

8. Basophils are the rarest of WBC, make .5-1% of all WBC, have 2 dissimilar lobes (or S shaped nucleus), and will be associated with allergies and radiation. The anticoagulant heparin is an important chemical produced.

9. Eosinophils comprise 2-4% of all WBC, are identified by 2 identical lobes, are very grainy, help to prevent the spread of inflammatory chemicals and destroy parasites.

10. The granulocytes are neutrophil, basophil and eosinophil.

11. Lymphocytes are the second most common WBC, make 20-25% of all WBC, are the smallest WBC, don't have a lobed nucleus, migrate to lymphatic tissues, some develop into B cells which produce antibodies, and some develop into T cells which fight

viruses and cancers.

12. Monocytes are the largest of all WBC, develop into macrophages, comprise 3-8% of all WBC, have a round or horseshoe shaped nucleus, and are associated with chronic infections.

13. The agranulocytes are the lymphocytes and monocytes.

14. Leukopoiesis is the production of WBC.

15. Leukopenia is an unusually low number of WBC.

16. Leukocytosis is an unusually high number of WBC.

THROMBOCYTES

Platelets (thrombocytes) are not cells they are only fragments of cells. These fragments come from large cells in the bone marrow called megakaryocytes. Imagine if you could grab the edges of the megakaryocyte and pinch it off a little at a time. The resulting pieces would be the platelets. Platelets are very important when it comes to preventing blood loss. They are always forming plugs over tiny microscopic tears in the tissues and also release chemicals needed for clotting.

Platelets are normally found in numbers between 250,000 and 400,000. These numbers are far less than RBC and far more numerous than WBC.

Platelets are actively involved with preventing the loss of blood. Preventing blood loss is also called hemostasis. Hemostasis involves three processes: vascular spasm, platelet plug formation and coagulation.

Vascular spasm is the constriction of the smooth muscle around blood vessels. The nervous system uses this muscle to regulate blood flow around the body. You may remember this from discussions on the sympathetic and parasympathetic divisions of the nervous system. When blood loss occurs, chemicals such as thromboxanes are responsible for the constriction. If the blood vessels constricts then less blood loss will occur.

Platelet plug formation is something that occurs every day in our body. As we go about our business each day, we are always causing microscopic tears in our tissues. As these tears occur, the inner lining of the blood vessels (endothelium) is torn and the connective tissue it was protecting is suddenly exposed to the blood. The platelets in the blood will now be exposed to collagen fibers in the connective tissue. The platelets will bind to the collagen when coming in contact with them. After binding to the collagen, platelets will release chemicals like ADP and thromboxanes. These chemicals will cause nearby platelets to stick to the collagen bound platelets. In addition the chemicals will cause more vasoconstriction. All of this platelet aggregation will form a plug over the tear, preventing blood loss. The steps of platelet plug formation are platelet adhesion, platelet release and aggregation.

Platelet facts to remember.

1. Platelets aren't cells, only fragments of cells.

2. Platelets are formed in the red bone marrow from megakaryocytes.

3. Normal platelet numbers are 250,000 – 400,000 per cubic millimeter.

4. Three processes occur to prevent blood loss: vascular spasm, platelet plug formation and coagulation.

5. Thrombocytopenia is a lower than normal platelet number.

6. Thrombopoiesis is the production of platelets.

COAGULATION

Coagulation is what we commonly call blood clotting. The formation of a blood clot is a complex series of chemical reactions, not fully understood. No less than 13 factors (different materials) have been identified in clot formation. Hopefully you won't have to know all of them. The formation of a clot can be broken down into a few important stages. Stage 1 is activation of prothrombinase. Prothrombinase is the enzyme that catalyzes prothrombin. Stage 2 is the conversion of prothrombin into thrombin. You may have heard of our need for vitamin K for clot formation. The vitamin K is needed for prothrombin function. Stage 3 is the conversion of fibrinogen to fibrin. The fibrin looks almost like a net within the blood clot. This net of fibers will hold back the formed elements of the blood and many other materials at the same time, thus preventing blood loss. You may also remember that adequate levels of calcium are required in the blood if it is going to clot.

Sometimes a clot will form when they are not needed and can be deadly. An attached clot is called a thrombus. Sometimes these attached clots will break free and floats in the blood. At this time the clot will become an embolus. An embolus may flow through the circulation until it reaches a blood vessel it can't pass through. The clot might stop blood flow and cause the death of tissues.

Hemophilia is a disorder where an individual lacks the gene needed to produce a protein needed for coagulation. Hemophilia A is caused by a lack of factor 8 and is the most common. Hemophilia B is a lack of factor 9.

PLASMA

Plasma is the watery part of the blood and normally comprises 55% of the blood. The plasma is 91-92% water at most times. Dissolved in this water are a large number of solutes. Proteins are one of the most common solutes in the plasma. Albumins are synthesized in the liver and work to buffer the blood. This protein makes up about 58% of the plasma proteins, so it has a large influence on blood viscosity and water balance. Globulins are the second most common plasma protein, making up 38% of plasma proteins. The globulins are largely antibodies, important components of the immune system. The last plasma protein is the fibrinogen, making up about 4% of the plasma proteins. Fibrinogen is essential to the clotting of blood.

Other solutes in the plasma include: nutrients, wastes, gases, hormones and ions. Nutrients are anything cells need for normal function and will include water, sugar, amino acids and many other substances. Wastes are materials produced by the cell, which must be removed to maintain homeostasis. Gases include oxygen, carbon dioxide and other gases. Ions are the charged particles responsible for anything from membrane potentials, action potentials, water balance, pH balance, etc. Hormones are regulatory chemicals produced by many organs of the body.

BLOOD TYPES

Blood types have been identified for over 100 years and dozens of blood types are known worldwide. We will look at the major blood types and the characteristics of each. Before discussing the blood types, we need to define a few terms.

Antigens – Antigens will usually be defined as any substance which stimulates an immune response. In blood typing an antigen is a protein found in the plasma membrane of a RBC and it is used to identify the RBC.

Antibodies – Antibodies are proteins produced in response to an antigen. Antibodies will be proteins found in the plasma. If an antibody comes in contact with the antigen, it is specific to, agglutination will occur. Agglutination will result in the destruction of whatever cell the antigen is part of. This destruction will occur by the lysing (rupturing) of the cell.

Agglutination – The clumping of blood which occurs when antigens and antibodies combine. This is not the same thing as coagulation.

The common blood typing system is called the ABO system. The system was given this name because most people have the blood type A, B, AB or O. Something added to these letters is what's called the Rh factor. The Rh factor is also called the D antigen and is what will give a blood type the positive or negative designation.

A person's blood type is usually determined at the time of a transfusion or donation. When a person is to receive blood, we want a proper matching of blood types. Mixing different blood types could end in death.

Make sure you understand the ABO blood types. The following table lists the blood type, antigens and antibodies you need to know.

Blood type	Antigens	Antibodies
A	A	anti-b
B	B	anti-a
AB	A and B	neither
O	neither	anti-a and anti-b

So if a person has type A blood, then they have A antigens on their RBC and anti-b antibodies in their plasma.

If a person has type B blood, they have B antigens on their RBC and anti-a antibodies in their plasma.

If a person has type AB blood, they have A and B antigens on their RBC, and neither antibody will be found in the plasma.

If a person has type O blood, they have neither antigen on their RBC, but they have both antibodies in their plasma.

What will be added to these blood types is the Rh factor (D antigen). This Rh factor is another antigen found on the RBC. If a person has this Rh factor, you simply put a + by their blood type. If a person doesn't have the Rh factor, you simply put a – by their blood type. A positive indicates the presence of the Rh protein and a negative indicates the absence of the protein.

So if a person has their blood typed and the following antigens are found: A, B and the Rh factor. What would the blood type be? AB+.

If a person has their blood typed and the following is found: no antigens present. What would the blood type be? O-.

If a person has their blood typed and the following antigens are found: A. The blood type would be A-.

The most common blood type is O+ followed closely by A+. The others are a bit rarer and every text will have different numbers when discussing what percent of the population has each type.

Transfusion reactions are what will occur if different blood types are mixed. If a person with type A blood is given blood from a type B donor, the result could be deadly. The anti-a antibodies of the donor are going to destroy the RBC of the recipient and the anti-b antibodies of the recipient are going to destroy the RBC of the donor blood. Remember that when antigens and antibodies of the same type come in contact with each other, RBC destruction occurs. So we only want to give a person their own blood type in an ideal situation. Some blood types can be mixed and there usually won't be a problem. But you ask, "What about type O blood being the universal donor?" If you look at type O blood, you will see it doesn't have any antigens on the surface of the RBC, so you don't have to worry about the type O RBC reacting with any antibodies from any other recipient. But notice that type O blood has both antibodies. If these antibodies are given to any other blood type, these antibodies will react with the antigens on the RBC of the recipient. This would result in the destruction of RBC. Getting around this problem is not difficult. If we take type O blood and centrifuge (spin at high speed) the blood, we can separate the cells from the plasma. Remember the antibodies are in the plasma, so if we separate cells from plasma by centrifuging we can remove the plasma, which will remove the antibodies. Any of the other blood types can receive the RBC of a type O donor, but you don't want to give them the plasma of the type O blood.

Chapter 2 – Study Questions

1. Blood is what type of tissue?
a. epithelial
b. connective
c. muscular
d. nervous
e. none of the above

2. The formed elements are the _____ of the blood?
a. solutes
b. ions
c. proteins
d. cells
e. all of the above

3. Red blood cells are also called?
a. erythrocytes
b. leukocytes
c. thrombocytes
d. platelets
e. none of the above

4. Red blood cells normally make what percent of total blood volume?
a. 30%
b. 45%
c. 55%
d. 65%
e. 75%

5. The percent that red blood cells make of total blood volume is?
a. erythrocytosis
b. leukopenia
c. hematocrit
d. antigens
e. antibodies

6. Red blood cells must be thick around the edges and thin in the center. This shape is called?
   a. obtuse
   b. circular
   c. flattened
   d. biconcave
   e. convex

7. Why do red blood cells need to be thin in the center and thick around the edges?
   a. for proper production of hemoglobin
   b. for flexibility
   c. to keep their red color
   d. to protect the hemoglobin
   e. none of the above

8. Sickle cell anemia is an adaptation to?
   a. hot temperatures
   b. cold temperatures
   c. malaria
   d. oxygen transport
   e. carbon dioxide

9. The hormone erythropoietin is released from what organ?
   a. brain
   b. liver
   c. kidneys
   d. pancreas
   e. thyroid

10. Erythropoietin has a target tissue of?
    a. heart
    b. arteries
    c. red bone marrow
    d. white blood cells
    e. red blood cells

11. Males should have how many RBC per cubic

millimeter?
- a. 7000
- b. 5.5 million
- c. 4.8 million
- d. 250,000
- e. 11000

12. A normal size for a RBC is how many microns in diameter?
- a. 3.0
- b. 5.5
- c. 7.0
- d. 7.5
- e. 10.0

13. The primary function of a RBC is?
- a. fight disease
- b. clot formation
- c. platelet plug formation
- d. destroy parasites
- e. transport oxygen

14. What element is needed for hemoglobin production?
- a. iron
- b. oxygen
- c. hydrogen
- d. chloride
- e. sodium

15. How long does a red blood cell live?
- a. 6 days
- b. 50 days
- c. 110 days
- d. 3 months
- e. years

16. We lose about what percent of our red blood cells each day?

a. 1
b. 5
c. 45
d. 55
e. 100

17. If a large number of RBC is being lost, what condition might develop?
a. polycythemia
b. viral infection
c. jaundice
d. leukopenia
e. leukocytosis

18. An excess of RBC is?
a. polycythemia
b. anemia
c. jaundice
d. leukopenia
e. leukocytosis

19. WBC are also called?
a. erythrocytes
b. leukocytes
c. thrombocytes
d. platelets
e. none of the above

20. How many major types of WBC exist?
a. 3
b. 5
c. 8
d. 10
e. 100

21. Which of the following is not a granulocyte?
a. neutrophil
b. eosinophil
c. basophil

d. lymphocyte
e. all are granulocytes

22. Macrophages used to be what earlier in their life?
a. neutrophil
b. eosinophil
c. monocyte
d. lymphocyte
e. basophil

23. The process by which WBC leave the blood and enter the tissues is?
a. chemotaxis
b. diapedesis
c. leukopoiesis
d. thrombopoiesis
e. anemia

24. The process by which WBC follow chemical trails to the source?
a. chemotaxis
b. diapedesis
c. leukopoiesis
d. thrombopoiesis
e. anemia

25. The most common of WBC is?
a. neutrophil
b. eosinophil
c. basophil
d. lymphocyte
e. monocyte

26. The smallest of all WBC is?
a. neutrophil
b. eosinophil
c. basophil
d. lymphocyte

e. monocyte

27. The largest of all WBC is?
a. neutrophil
b. eosinophil
c. basophil
d. lymphocyte
e. monocyte

28. Which WBC might have 5 lobes to its nucleus?
a. neutrophil
b. eosinophil
c. basophil
d. lymphocyte
e. monocyte

29. Which WBC will have 2 identical lobes?
a. neutrophil
b. eosinophil
c. basophil
d. lymphocyte
e. monocyte

30. Which WBC is the rarest?
a. neutrophil
b. eosinophil
c. basophil
d. lymphocyte
e. monocyte

31. Which WBC would be elevated during acute bacterial infections?
a. neutrophil
b. eosinophil
c. basophil
d. lymphocyte
e. monocyte

32. Which WBC are elevated after radiation exposure?
a. neutrophil
b. eosinophil
c. basophil
d. lymphocyte
e. monocyte

33. Which WBC might have an S shaped nucleus?
a. neutrophil
b. eosinophil
c. basophil
d. lymphocyte
e. monocyte

34. Which WBC can release heparin?
a. neutrophil
b. eosinophil
c. basophil
d. lymphocyte
e. monocyte

35. Which WBC would be elevated during a parasite infection?
a. neutrophil
b. eosinophil
c. basophil
d. lymphocyte
e. monocyte

36. Which WBC will quickly move to the lymphatic system?
a. neutrophil
b. eosinophil
c. basophil
d. lymphocyte
e. monocyte

37. Which WBC will differentiate into B and T cells?
a. neutrophil

b. eosinophil
c. basophil
d. lymphocyte
e. monocyte

38. An unusually low number of WBC is?
a. polycythemia
b. anemia
c. jaundice
d. leukopenia
e. leukocytosis

39. An unusually high number of WBC is?
a. polycythemia
b. anemia
c. jaundice
d. leukopenia
e. leukocytosis

40. Platelets are also known as?
a. erythrocytes
b. leukocytes
c. thrombocytes
d. all of the above
e. none of the above

41. Which of the following are cell fragments?
a. erythrocytes
b. leukocytes
c. thrombocytes
d. all of the above
e. none of the above

42. Normal numbers for platelets per cubic millimeter are?
a. 5.5 million
b. 7000
c. 4.8 million
d. 300,000

e. 1 million

43. Platelets are involved with?
a. fight disease
b. radiation sickness
c. platelet plug formation
d. destroy parasites
e. transport oxygen

44. A lower than normal number of platelets is?
a. polycythemia
b. anemia
c. jaundice
d. leukopenia
e. thrombocytopenia

45. What vitamin is needed for coagulation?
a. A
b. B
c. C
d. D
e. K

46. An attached clot is called a?
a. thrombus
b. embolus
c. hemophilia
d. anemia
e. thrombin

47. A clot free and floating in the blood is?
a. thrombus
b. embolus
c. hemophilia
d. anemia
e. thrombin

48. Hemophilia A is caused by a lack of factor?

a. 5
b. 6
c. 7
d. 8
e. 9

49. Plasma should make what % of the blood?
a. 30%
b. 45%
c. 55%
d. 65%
e. 75%

50. The most common protein in the plasma is?
a. albumin
b. globulin
c. fibrinogen

51. The most common blood type is?
a. A+
b. B-
c. AB+
d. O+

52. Blood types are identified by proteins on the cell surface. What is the name of these proteins?
a. antigens
b. antibodies
c. factors
d. agglutinations
e. ions

53. A person with type A blood will have what antigens?
a. A
b. B
c. AB
d. D
e. none

54. A person with type O blood will have what antigens?
a. A
b. B
c. AB
d. D
e. none

55. A person with type B blood will have what antibodies?
a. anti A
b. anti B
c. both
d. neither

56. A person with type O blood will have what antibodies?
a. anti A
b. anti B
c. both
d. neither

57. The Rh factor is also called the?
a. A antigen
b. B antigen
c. D antigen
d. antibody

Answers to multiple choice questions.
1. B
2. D
3. A
4. B
5. C
6. D
7. B
8. C
9. C
10. C
11. B
12. D
13. E
14. A
15. C
16. A
17. C
18. A
19. B
20. B
21. D
22. C
23. B
24. A
25. A
26. D
27. E
28. A
29. B
30. C
31. A
32. C
33. C
34. C
35. B
36. D
37. D

38. D
39. E
40. C
41. C
42. D
43. C
44. E
45. E
46. A
47. B
48. D
49. C
50. A
51. D
52. A
53. A
54. E
55. A
56. C
57. C

HEART

The pathway of blood flow through the heart

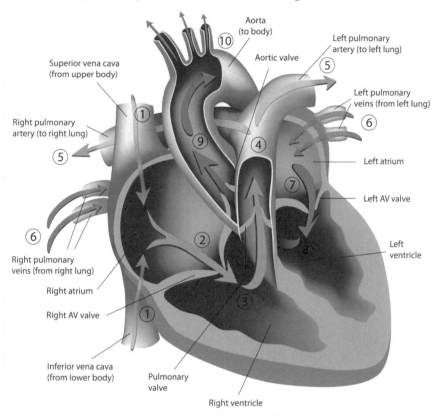

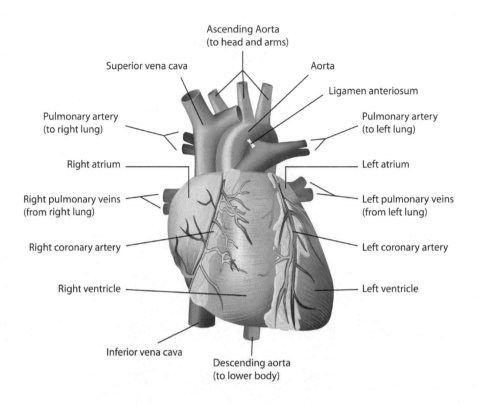

ANTERIOR HEART ANATOMY

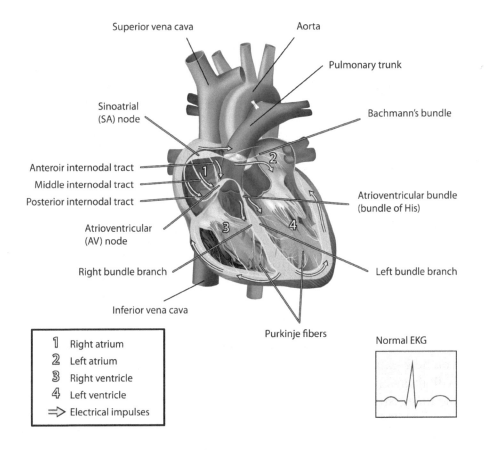

Electric Conducting System of the Heart

1. SA node-sinoatrial node
2. AV node-atrioventricular node
3. AV bundle-atrioventricular bundle (bundle of His)
4. 2 bundle branches
5. Purkinje fibers

CHAPTER 3
HEART

The heart belongs to the cardiovascular system. Along with the heart we will find arteries, veins and blood in the cardiovascular system. We often think of the heart as a pump, but you need to think of it as two pumps. We will soon see that the heart has a pump on the right side and the left side. The right pump is receiving blood from the body and pumping it to the lungs. The left pump is receiving blood from the lungs and pumping it to the body. Notice how these two pumps receive and pump blood to the opposite structures.

One of the first things you need to change in this chapter is how you define the heart. Get in the habit of saying, "Right pump and left pump", when you refer to the heart.

The right pump is all about pulmonary circulation. When you hear pulmonary, think about the lungs, this is what pulmonary will always have to do with. The right pump is pumping blood to the lungs, so that the blood can release carbon dioxide and pick up oxygen. This is primarily why we have lungs. We must pick up the oxygen for aerobic respiration and we must release the carbon dioxide to maintain pH balance.

The left pump is all about systemic circulation. When you hear systemic, think about all of the body, except the lungs. This pump receives blood from the lungs and blood coming from the lungs will be oxygen rich. The left pump is pumping blood out to the body, so cells can receive the oxygen they need.

The heart is located in the thoracic cavity, within the region between our lungs called the mediastinum. Within the mediastinum we find the pericardial cavity and our heart is in it. Within the thoracic cavity our heart is protected by the hard bones of the sternum and ribs, which surround it. Our heart is found more on the left side of our chest. About 2/3 of our heart is on the left and this is why our left lung is smaller than the right one. You will notice in lung illustrations, that the left lung has two lobes and the right lung has three.

The heart contains four chambers and four valves. The superior surface of the heart is the broad base. From the base there are several large arteries and veins. The inferior part of the heart comes to a point and is called the apex.

PERICARDIAL CAVITY

The pericardial cavity has within it two serous membranes. The inner membrane is the visceral pericardium and the outer membrane is the parietal pericardium. The visceral layer is the outer surface of the heart. The parietal layer is surrounding it, like a superficial skin. The two layers have between them pericardial fluid. These layers and fluid always serve two purposes: reduce friction and hold the organ in place. To imagine how this works, imagine two pieces of glass, one on top of another. If the two pieces of glass are dry and you push one across the other, they will scratch each other. If you take the same two pieces of glass and put a little water in between them, they slide gently over each other. This is how the membranes and fluid reduce friction. Take the same two pieces of glass when dry and you can easily separate them. Put the water back between them and try to pull them apart, and it becomes difficult. This is how the membranes and fluid hold the heart in place.

Sometimes the layers of the pericardial cavity can become damaged. If damaged the tissues will become inflamed, this is called pericarditis. If fluid collects within the pericardial sac, the condition is now called cardiac tamponade. Cardiac tamponade causes big problems with the heart. If fluid collects around the heart, this will put pressure on the heart. Remember that muscle contracts forcefully, but it relaxes passively, meaning that muscle needs something else to return it to resting length. If the heart has pressure around it, it can't refill with blood. If it can't refill with blood, it can't pump blood.

HEART LAYERS

The heart has three layers and if you are familiar with the prefixes epi-, endo- and myo-, you will easily understand these layers. The meaning of epi is outer, endo is inner and myo means muscle. The outer layer of the heart is the epicardium, the inner is the endocardium and the middle layer is the myocardium. The middle layer is by far the thickest and is where the cardiac muscle is found. The inner layer is very smooth so as to cause as little friction as possible against the blood flow.

HEART ANATOMY

ATRIA

The heart is separated into two upper atria and two lower ventricles. The upper chambers are called atria and consist of a right and left. Remember when thinking right and left, always thing about your right and left. These upper chambers are small primer pumps for the large ventricles inferior to them. These upper atria will pump blood down through atrioventricular valves to the ventricles. The two atria are separated by the thin interatrial septum. A septum will always be a separating structure. The veins leading back to the heart will bring blood into these atria. The right atrium will receive blood from three sources: the superior vena cava, the inferior vena cava and the coronary sinus. The superior vena cava returns blood from above the heart, the inferior vena cava returns blood from below the heart and the coronary sinus returns blood from the cardiac muscle itself. The left atrium receives blood from the four pulmonary veins, which return blood from the lungs. On the superficial surface of the atria are flap like structures called auricles. The auricles look like ear flaps on the atria.

VENTRICLES

The two lower chambers are the ventricles and also consist of a right and left. The ventricles are the large power pumps found on bottom. These are the chambers which move blood where it is needed. The right ventricle pumps blood to the lungs and the left ventricle pumps blood to the rest of the body. The ventricles are separated by the thick interventricular septum. The ventricles receive blood from the upper atria, which only work to fill the ventricles with extra blood. The more fully the ventricles are filled, the more blood they can pump. The ventricles are much thicker in cross section than the atria, because they have much more muscle in them than the atria do. Also the left ventricle has much more muscle in it than the right ventricle. The left ventricle has to push blood from the top of the head to the tip or our toes. It takes a large amount of pressure to push blood that far, so the left ventricle has a large amount of muscle in it. The right ventricle pushes blood to the lungs, which is close to the heart. Pushing blood a small distance is easy, so the right ventricle has less muscle than the left ventricle. Atria only have to push blood a short distance (down to the ventricles), this takes a small amount of muscle. The ventricles must push blood much further. Pushing blood a greater distance, requires much more force, thus more muscle is needed.

VALVES

The heart contains four valves. The two atrioventricular valves are found in between the atria and ventricles. The right atrioventricular valve is also called the tricuspid valve, because of the three cusps it possesses. Think of each cusp as like a parachute. When air catches in the bottom of a parachute, the parachute will fill and catch the air. When blood pushes against these valves, they will catch the blood and prevent its backflow. Valves are always there to prevent backflow. The left atrioventricular valve is also called the bicuspid or mitral valve. Extending from these AV valves are rope like structures called chordae tendineae (heart strings). These strings are attached to the papillary muscles, which are extensions of the ventricles. The papillary muscles will pull on the chordae tendineae, when the ventricles contract. This pulling keeps tension on the valves and prevents them from prolapsing (falling back) into the atria. Think of this tension on these strings as being like the tension a person applies on a parachute, when they are falling back to earth. The right AV valve will prevent back flow into the right atrium and the left AV valve will prevent back flow into the left atrium.

The semilunar valves are found just inside the aorta and pulmonary trunk. The pulmonary semilunar valve is between the right ventricle and the pulmonary trunk. This valve will prevent back flow from the lungs. The aortic semilunar valve is between the left ventricle and the aorta. This valve will prevent back flow from the body. All of these valves ensure a proper one way flow to the blood.

If you want to learn most of the structures of the heart, the best way to do this is to follow a drop of blood through the heart. Look at a picture of the interior of the heart, while reading these 15 steps and you will know much of the heart anatomy in no time.

15 STEPS TO BLOOD FLOW THROUGH THE HEART

Start with the three inflows on the right side of the heart.

1. Superior vena cava, inferior vena cava and coronary sinus.
2. Right Atrium
3. Right atrioventricular valve (tricuspid valve)
4. Right ventricle
5. Pulmonary semilunar valve
6. Pulmonary trunk
7. Pulmonary arteries
8. Lungs
9. Pulmonary veins
10. Left atrium
11. Left atrioventricular valve (bicuspid, mitral)
12. Left ventricle
13. Aortic semilunar valve
14. Aorta
15. Out to the body

CIRCULATION TO THE MYOCARDIUM

On the anterior and posterior surface of the heart, you will find several large arteries taking blood to the cardiac muscle and several large veins returning blood from the cardiac muscle. The major arteries supplying the heart with blood are the two coronary arteries, two interventricular arteries and the two marginal arteries. Blockage of any of these arteries can quickly result in inadequate oxygen delivery to the myocardium. The result could be a myocardial infarction (heart attack). The three major veins taking blood from the cardiac muscle are the great cardiac vein, small cardiac vein and middle cardiac vein. These three will return blood to the coronary sinus, which leads to the right atrium.

SKELETON OF THE HEART

The heart doesn't have a skeleton of bone, but one of connective tissue. This tissue will always serve three functions. First it will serve as an insulator to prevent action potentials from traveling an undesired path. Second it provides a strong anchor for the cardiac muscle cells. Lastly it will help to keep the area around the valves open to ensure blood flow.

ELECTRIC CONDUCTING SYSTEM OF THE HEART

The conducting system has five components. Make sure you are aware of the order of all five components, the location of each and how they work together. The conducting system is a series of specialized cardiac muscle cells, which ensure the proper timing and direction of action potentials through the heart.

Electric Conducting System of Heart in order

a. SA node – sinoatrial node (pacemaker)
b. AV node
c. AV bundle
d. 2 bundle branches
e. purkinje fibers

The first part of the electric conducting system is the SA node. The SA node is located in the superior part of the right atrium. The SA node is also called the pacemaker of the heart, because this is where all action potentials in the heart should begin. When the SA node generates an action potential, the electric signals will spread rapidly across the atria. This will ensure that the two atria contract together. This will ensure that the atria (primer pumps) work to fill the ventricles as fully as possible before the ventricles contract.

As the action potentials spread across the right atrium the action potentials will reach the AV node, located in the inferior region of the right atrium. The action potentials will leave the AV node and travel to the AV bundle. When traveling to the AV bundle the action potentials will slow. This slowing gives the atria time to complete contraction, before the action potential travels to the ventricles. If the atria didn't have this time the ventricles wouldn't be properly filled and stroke volume would decrease.

After leaving the AV bundle the action potentials will travel rapidly down the two bundle branches. The bundle branches are located within the interventricular septum. These action potentials will travel rapidly to the apex of the heart and then spread out through the purkinje fibers in the ventricles. Contraction of the ventricles always starts at the apex of the heart, because this is where the ventricles are first stimulated.

Sometimes some part of the heart other than the SA node begins the action potential in the electric conducting system. This is called an ectopic pacemaker and will lead to the heart beating in an arrhythmic fashion.

ELECTROCARDIOGRAM (ECG)

If you think about the name electrocardiogram, you see it has something to do with electric signals and the heart. An ECG will receive and monitor electric signals generated from the heart. We can relate these electric events to mechanical events in the heart. You need to know what a normal ECG looks like and what the three waves on it are telling us. The ECG illustrates the sequence of events occurring in the heart, how strong they are and how long each wave takes. Each square on an ECG represents .04 seconds, so it is easy to count squares to see how long a wave takes. The normal ECG has three waves: P wave, QRS complex and the T wave.

The P wave shows when and for how long the atria are depolarizing. When the atria are depolarizing they will be contracting. Recall from the electric conducting system of the heart, that the atria should contract first and in a normal ECG this can be seen. The PQ interval is the time from the beginning of the P wave to the beginning of the Q wave. This PQ interval should occupy four squares on the ECG, so it should take .16 seconds for the atria to complete depolarization. Consider the problem if this interval occupied only three squares on the ECG. If the atria don't have time to complete depolarization and contraction the ventricles won't be fully filled.

The second wave is the QRS complex and is a much stronger wave than the first. This complex represents depolarization of the ventricles, which means contraction of the ventricles will occur at this time. Since the ventricles contain far more muscle cells than the atria, more action potentials are generated at this time than what is generated during the P wave. With more action potentials being generated, the wave will be stronger. The atria repolarize at the same time the ventricles are depolarizing. Ordinarily this repolarization would produce a wave, but the depolarization of the ventricles over shadows the smaller wave generated by the repolarizing atria. So remember this wave (QRS) represents ventricular depolarization and atrial repolarization. The QRS complex should take about .08 seconds.

The third wave is the T wave and represents repolarization of the ventricles. The ventricles should take about .36 seconds to contract and relax. This .36 seconds would be the QT interval.

CARDIAC CYCLE

The cardiac cycle includes all of the events occurring in the heart during one complete cycle of contraction and relaxation of all chambers. The cycle is broken down into systole (contraction) and diastole (relaxation). What the cardiac cycle shows us is the rhythmic functioning of the chambers and valves. When looking at the cardiac cycle, notice how the sets of chambers work opposite of each other. When the atria are contracting the ventricles are relaxing. When the ventricles are contracting the atria are relaxing. You will notice the same about the sets of valves. When the AV valves are closed the semilunar valves are open. When the semilunar valves are closed the AV valves are open.

Step 1 of the cardiac cycle is isovolumetric ventricular contraction. During this step the ventricles are contracting, but haven't yet produced enough pressure to overcome the back flow pressure against the semilunar valves. The AV valves are closed to prevent back flow into the atria.

Step 2 of the cardiac cycle is ejection. During this step the ventricles have generated enough pressure to move blood from the ventricles, through the semilunar valves and out the aorta and pulmonary trunk. These first two steps are systole. A pressure gradient has been created and blood moves from high pressure in the ventricles to low pressure in the aorta and pulmonary trunk. The first two steps represent systole.

Step 3 of the cardiac cycle is isovolumetric ventricular relaxation. The ventricles have started to relax and this will cause the semilunar valves to close as blood tries to back flow.

Step 4 of the cardiac cycle is passive ventricular filling. During passive filling the ventricles continue to relax and the AV valves open as pressure in the ventricles drops. Blood moves from the atria to the ventricles and this is when the ventricle receives most of its blood.

Step 5 of the cardiac cycle is active ventricular filling. During active filling the atria contract and complete the filling of the ventricles. The last three steps are diastole.

The entire process repeats itself with each contraction.

HEART SOUNDS

We all know the characteristic "bump bump" sound of the heart. The sounds are not the sound of the chambers contracting, but the sounds of the valves slamming shut. Think about which valves would close first. Looking at the cardiac cycle we see that the ventricles contract in the first part of the cardiac cycle (systole). What valves would be closing at this time? Which valves prevent back flow during ventricular systole? The atrioventricular valves prevent this back flow. The slamming of the AV valves is responsible for the first heart sound.

The ventricles relax in the second part of the cardiac cycle. As the ventricles relax, which valves prevent back flow into the ventricles? The two semilunar valves close during diastole to prevent back flow. The two semilunar valves closing are responsible for the second heart sound.

CARDIAC OUTPUT

Cardiac output is also called minute volume, because it is the volume of blood pumped in one minute by the heart. If you want to know how much blood the heart pumps in one minute, you only need to know two variables; how many times did the heart beat in a minute and how much blood did it pump with each contraction. Heart rate is measured in beats per minute and the stroke volume is how much blood is pumped with each contraction.

The average adult has a resting heart rate around 72 bpm and will pump about 70 ml of blood with each contraction. If we multiply 72 X 70 we get about 5 liters of blood per minute. This would be the cardiac output for the average person when resting.

When a person is exercising the sympathetic division of the nervous system will do two things to our heart. It will make the heart beat faster and with more forceful contractions. If the heart beats faster, then our cardiac output will increase. If the heart contracts more forcefully then it will push more blood with each contraction. If we increase our heart rate and make the muscle work more forcefully, then we will greatly increase our cardiac output. When exercising our heart rate may increase to about 190 bpm and the stroke volume might increase to about 115 ml. This would give a cardiac output of about 22 liters per minute. That is a large increase from 5 to 22 liters. The difference between the cardiac output at rest and the cardiac output while exercising is called the cardiac reserve. The greater the difference in these two numbers, the greater the capacity a person has to do additional work.

There are a few factors which affect stroke volume of the heart. One of these is the extent to which the cardiac muscle cells are stretched during relaxation. This was first described by Ernest Starling and is now called Starling's Law (also called preload). Starling's Law states that the greater the extent to which the cardiac cells are stretched, the greater the stroke volume will be. This makes sense if you think about what is happening at the time the cells are being stretched. When the cardiac cells are being stretched, this is the time in which the ventricles are filling. The more the cells are stretched the greater the volume of blood that is filling the ventricles. The more the ventricles are filled, the more blood they will pump, thus increasing stroke volume.

You can also think of Starling's Law as not just, how much the cardiac cells are being stretched, but also how much the ventricles are being filled. The stretching and the filling are the same thing. It makes sense that if the ventricles are more fully filled, they will push more blood with each contraction. If more blood is pushed with each contraction, stroke volume increases. Starling's Law has a large effect on stroke volume.

What is largely responsible for increasing stroke volume is the contraction of skeletal muscles, when we are exercising. When we exercise, we contract skeletal muscles. Skeletal muscles hold a large volume of blood when we are exercising. As the skeletal muscles contract, they squeeze the blood inside of them. The blood is squeezed back towards the heart through the veins and the valves in the veins will prevent back flow. This squeezing of blood back to the heart will fill the ventricles more fully. When the ventricles are filled more fully, the ventricles will push more blood. If the heart is going to pump more blood, then more blood must be returned to it.

Another factor affecting the stroke volume is afterload. Think of afterload as the back flow pressure against the semilunar valves. The pressure against the pulmonary semilunar valve is not that great, we won't worry about it. The pressure against the aortic semilunar valve is much greater. This valve is preventing back flow from most all of the body. The left ventricle must overcome this back flow pressure to eject blood. The left ventricle is so strong that this back flow pressure has a small effect on stroke volume.

SYMPATHETIC AND PARASYMPATHETIC CONTROL

The sympathetic division of the nervous system has a large effect on the heart. We can see a resting heart rate increase from 72 bpm up to around 300 bpm during intense sympathetic stimulation. This is accomplished by the release of epinephrine and norepinephrine.

The parasympathetic division has a small effect on the heart. The parasympathetic division can decrease heart rate by about 30 bpm. Obviously this is a small change in comparison to the sympathetic.

HEART RATE REGULATION

Ever wonder, "How does our body know what our heart rate and blood pressure are?" The body monitors blood pressure and heart rate through the use of baroreceptors and chemoreceptors.

A barometer detects changes in pressure, so it makes sense that baroreceptors would monitor blood pressure. We have these baroreceptors in a few large arteries of the body like the aorta and the internal carotids. Ask yourself, "Why would we want baroreceptors here?" The aorta is the first artery of the body, if we know what is happening with blood pressure there, then we know what will be happening to blood pressure everywhere. The internal carotids are the two largest arteries supplying blood to our brain. We must maintain blood pressure to the brain or we will be dead quickly. So these are good places to monitor blood pressure.

Baroreceptors can also be thought of as stretch receptors. They are monitoring pressure, but it is the stretching of the artery that influences these receptors. If you want to find elastic tissue in the body, arteries are a great place to find it. Because of this elasticity, they will stretch as the pressure inside of them increases.

When we are at rest our heart is not beating very fast or forcefully, so only a small amount of blood is being pushed into the arteries. When a small amount of blood is being pushed into the arteries, the arteries are stretched a small amount. A small amount of stretching will cause a baroreceptor to send action potentials back to the brain (medulla oblongata) infrequently. The brain perceives a low frequency as low blood pressure. When we exercise, our heart rate increases and more blood is pushed into the arteries. The more blood that is pushed into the arteries, the more the arteries are stretched. The more the arteries are stretched, the more frequently they send action potentials back to the brain. The brain sees a high frequency as high blood pressure. Remember that action potentials work by the all or none principle, so our brain can't look at the strength of action potentials, it must look at frequency.

Chemoreceptors will also affect heart rate and blood pressure. Our body monitors important chemicals like oxygen and carbon dioxide. Of the two carbon dioxide is the stronger regulator by far. We often think that oxygen levels are what primarily regulate heart rate and ventilation. This is a common misconception. Oxygen levels in the blood must decline greatly, before we see change in heart rate and ventilation. A very tiny change in carbon dioxide levels is sufficient to change heart rate and ventilation. The body is so sensitive to carbon dioxide levels because the CO_2 is tied directly to pH (hydrogen ion levels). Our pH range for the body is

7.35 – 7.45, with an ideal value of 7.4. That means our body doesn't like fluctuations greater than plus or minus .05 in either direction. When carbon dioxide is released, it will chemically combine with water to make carbonic acid. Carbonic acid will break down into hydrogen ion and bicarbonate ion. $CO_2 + H_2O \leftrightarrow H_2CO_3 \leftrightarrow H + HCO_3$. Because this is a reversible chemical reaction and because carbon dioxide is on one side and hydrogen ion is on the other, what happens to one of these materials, will happen to the other. So when carbon dioxide levels increase, so will hydrogen ion levels. When carbon dioxide levels decrease, so will hydrogen ion levels. Our body is so sensitive to carbon dioxide levels because when carbon dioxide increases so does hydrogen ion. When hydrogen ion levels change, so does pH levels.

Why is our body so sensitive to pH balance? It is because of proteins. The proteins in our body are very sensitive to temperature and pH changes. If either of these two variables changes far from what is normal, the proteins in our body change shape. When the proteins change shape they will no longer be able to perform their jobs in the body. When these proteins can't do their job, we are taken away from homeostasis.

Consider a person who has a heart or lung problem. If the heart has become weak and doesn't pump blood adequately to the lungs, gas exchange can be inadequate. If a person has a lung problem, gas exchange can be inadequate. If gas exchange is inadequate, we might not get in enough oxygen and out enough carbon dioxide. To help with this gas exchange problem a person can be given additional oxygen, via a tube and oxygen tank. But how can we remove carbon dioxide from the blood? We can't. When carbon dioxide levels build, the hydrogen ion levels will build. This increase in hydrogen ions will cause acidosis (low pH number). When the pH changes the proteins in our body start to change shape, they no longer do their job and people die. There are ways to temporarily treat the pH changes, but the gas exchange problem must be fixed.

HEART DISORDERS

Myocardial infarction. To understand this disorder you need to break down the two terms. Myocardial refers to the cardiac muscle in the heart. An infarct is an area of cell death. This literally means a death of cardiac muscle. A myocardial infarction is what we commonly call a heart attack. If the cardiac muscle cells are deprived of oxygen for more than 20 minutes, there is a good chance that some of the cardiac cells will die. Once muscle cells are lost, they can't be replaced.

Angina pectoris. This refers to sudden and severe pain in the chest, often accompanied by a suffocating feeling. This is not the same as a heart attack, but could be a warning that the heart is not receiving enough oxygen.

Congestive heart failure. This comes in two forms: left side and right side congestive heart failure. Left side congestive heart failure occurs when the left pump (ventricle) is failing. Consider if the right pump is strong and the left pump is weak, "What would happen?" The right pump receives blood from the body and pushes it to the lungs. The left pump receives blood from the lungs and pushes it to the body. If the left pump becomes weak, then we have a strong pump pushing blood into the lungs and a weak pump trying to push it out. The end result will be an accumulation of fluid in the lungs (pulmonary edema). This fluid buildup will thicken the respiratory membrane and inhibit gas exchange.

Right side congestive heart failure is when the right pump is weak and failing. If the left pump is strong, then we have plenty of blood pumped out to the body, but if the right pump is weak it can't pump all of this blood out of the body. The result will be edema in all parts of the body except the lungs. Fluid accumulation will probably be seen in the lower limbs first and work its way back towards the heart.

In summary left side congestive heart failure causes fluid buildup in the lungs and right side congestive heart failure causes fluid buildup everywhere else.

Heart bypass surgery. In heart bypass surgery an artery supplying blood to the cardiac muscle has become blocked. A surgeon can take a vein from the lower limbs, usually the great saphenous, and attach it between the aorta and the part of the heart needing more blood. This will bypass any blockage and restore blood flow.

Bradycardia. A slower than normal heart rhythm. A heart rate less than 60 bpm is considered bradycardia.

Tachycardia. A faster than normal heart rhythm. A heart rate greater than 100 bpm is considered tachycardia.

Ventricular fibrillation. A loss of ventricular function.

Hypertension. Commonly called high blood pressure. Most texts will consider 120/80 normal blood pressure.

Chapter 3 – Study Questions

1. What is not part of the cardiovascular system?
a. blood
b. heart
c. arteries
d. veins
e. red bone marrow

2. The right pump of our heart receives blood from where?
a. body
b. lungs
c. brain
d. internal organs
e. upper and lower limbs

3. The right pump of our heart pumps blood to?
a. body
b. lungs
c. brain
d. internal organs
e. upper and lower limbs

4. The left pump of our heart receives blood from where?
a. body
b. lungs
c. brain
d. internal organs
e. upper and lower limbs

5. The left pump of our heart pumps blood to?
a. body
b. lungs
c. brain
d. internal organs
e. upper and lower limbs

6. The right pump is all about which circulation?
a. pulmonary

b. systemic

7. The left pump is all about which circulation?
a. pulmonary
b. systemic

8. Our heart lies more to what side of the thoracic cavity?
a. left
b. right
c. neither

9. The heart contains how many chambers?
a. 1
b. 2
c. 3
d. 4
e. 5

10. The pericardial cavity has 2 membranes, which one is the inner membrane?
a. visceral
b. parietal

11. What are the functions of the pericardial membranes and pericardial fluid?
a. reduce friction
b. hold the heart in place
c. both
d. neither

12. If fluid collects in the pericardial sac, a person will have what condition?
a. pleurisy
b. pericarditis
c. cardiac tamponade
d. thoracic inflammation
e. hypertension

13. Which heart layer will be found on the outer surface of the heart?
   a. epicardium
   b. endocardium
   c. myocardium

14. Which heart layer will be found on the inner surface of the heart?
   a. epicardium
   b. endocardium
   c. myocardium

15. Which heart layer is the thickest?
   a. epicardium
   b. endocardium
   c. myocardium

16. The upper chambers of the heart are the?
   a. ventricles
   b. atria

17. The lower chambers of the heart are the?
   a. ventricles
   b. atria

18. Which chambers of the heart are the largest and strongest?
   a. ventricles
   b. atria

19. The valve between the right atrium and right ventricle?
   a. pulmonary semilunar valve
   b. aortic semilunar valve
   c. tricuspid
   d. bicuspid

20. The valve between the left atrium and left ventricle?
   a. pulmonary semilunar valve

b. aortic semilunar valve
c. tricuspid
d. bicuspid

21. The mitral valve is also called the?
a. pulmonary semilunar valve
b. aortic semilunar valve
c. tricuspid
d. bicuspid

22. Which valve prevents back flow from the right ventricle?
a. pulmonary semilunar valve
b. aortic semilunar valve
c. tricuspid
d. bicuspid

23. Which valve prevents back flow from the left ventricle?
a. pulmonary semilunar valve
b. aortic semilunar valve
c. tricuspid
d. bicuspid

24. Which valve prevents back flow from the lungs?
a. pulmonary semilunar valve
b. aortic semilunar valve
c. tricuspid
d. bicuspid

25. Which valve prevents back flow from the body?
a. pulmonary semilunar valve
b. aortic semilunar valve
c. tricuspid
d. bicuspid

26. Blood from the pulmonary trunk will flow to what structure next?
a. right atrium
b. left atrium

c. body
d. pulmonary arteries
e. superior vena cave

27. Blood from the left atrium will flow to what structure next?
a. pulmonary arteries
b. pulmonary veins
c. aorta
d. lungs
e. bicuspid

28. The proper order for the electric conducting system of the heart is?
a. AV node, SA node, AV bundle, purkinje fibers, bundle branches
b. SA node, AV node, AV bundle, bundle branches, purkinje fibers
c. purkinje fibers, bundle branches, AV bundle, AV node, SA node
d. AV bundle, AV node, SA node, bundle branches, purkinje fibers
e. AV bundle, purkinje fibers, SA node, AV node, bundle branches

29. The pacemaker of the heart is the?
a. AV node
b. AV bundle
c. purkinje fibers
d. SA node
e. bundle branches

30. The P wave on an ECG represents?
a. atrial depolarization
b. ventricular depolarization
c. atrial repolarization
d. ventricular repolarization
e. all of the above

31. The QRS complex on an ECG represents?
a. atrial depolarization
b. ventricular depolarization
c. atrial contraction
d. ventricular repolarization
e. all of the above

32. The T wave on an ECG represents?
a. atrial depolarization
b. ventricular depolarization
c. atrial repolarization
d. ventricular repolarization
e. all of the above

33. Each square on an ECG represents how much time?
a. .01 seconds
b. .04 seconds
c. .10 seconds
d. 1 second
e. 2 seconds

34. Which wave will be the strongest in millivolts?
a. P wave
b. QRS complex
c. T wave
d. none of the above
e. all of the above

35. The term systole means?
a. filling
b. passing
c. contraction
d. relaxation
e. removal

36. The term diastole means?
a. filling

b. passing
c. contraction
d. relaxation
e. removal

37. Most of the blood enters the ventricles during what part of the cardiac cycle?
a. isovolumetric ventricular contraction
b. ejection
c. isovolumetric ventricular relaxation
d. passive ventricular filling
e. active ventricular filling

38. The first heart sound is caused by?
a. closing of the AV valves
b. closing of the semilunar valves
c. opening of the AV valves
d. closing of all heart valves
e. none of the above

39. The second heart sound is caused by?
a. closing of the AV valves
b. closing of the semilunar valves
c. opening of the AV valves
d. closing of all heart valves
e. none of the above

40. The average resting heart rate is how many bpm?
a. 60
b. 120
c. 80
d. 72
e. 100

41. Tachycardia is a heart rate above?
a. 60
b. 120
c. 80

d. 72
e. 100

42. Bradycardia is a heart rate below?
a. 60
b. 120
c. 80
d. 72
e. 100

43. The cardiac output for the average resting person is how many liters per minute?
a. 1
b. 2
c. 5
d. 8
e. 10

44. Cardiac output for the average exercising person is how many liters per minute?
a. 10
b. 22
c. 15
d. 88
e. 11

45. What chemical is the brain most sensitive to?
a. oxygen
b. carbon dioxide
c. bicarbonate ion
d. nitrogen
e. water

46. A heart attack is also called?
a. myocardial infarction
b. angina pectoris
c. congestive heart failure
d. bradycardia

e. tachycardia

47. Rapid heart rate is also called?
a. myocardial infarction
b. angina pectoris
c. congestive heart failure
d. bradycardia
e. tachycardia

48. Right side congestive heart failure will cause fluid buildup where?
a. heart
b. lungs
c. body
d. brain
e. all of the above

49. Left side congestive heart failure will cause fluid buildup where?
a. heart
b. lungs
c. body
d. brain
e. all of the above

Answers to multiple choice questions.
1. E
2. A
3. B
4. B
5. A
6. A
7. B
8. A
9. D
10. A
11. C
12. C
13. A
14. B
15. C
16. B
17. A
18. A
19. C
20. D
21. D
22. C
23. D
24. A
25. B
26. D
27. E
28. B
29. D
30. A
31. B
32. D
33. B
34. B
35. C
36. D
37. D

38. A
39. B
40. D
41. E
42. A
43. C
44. B
45. B
46. A
47. E
48. C
49. B

ARTERIES AND VEINS

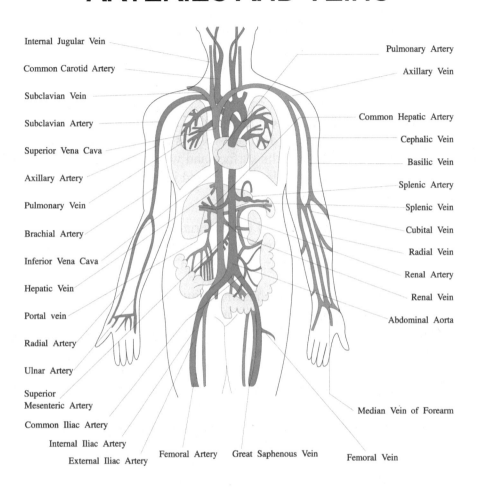

CHAPTER 4
ARTERIES AND VEINS

Our blood vessels are a pipeline system connecting almost all tissues of the body. Even though all tissues don't have blood vessels penetrating them, the tissues which don't will have blood vessels passing very close to them. The epidermis of our skin is an example of a tissue blood vessels don't penetrate. Why do you think blood vessels don't pass through the epidermis? The epidermis is a thin, superficial layer in our skin. If blood vessels passed through it, the blood would be too close to the surface and wouldn't have enough protection. We would lose blood easily and infection could enter the cardiovascular system easily.

Blood vessels deliver all the materials cells need to function properly and remove the wastes harmful to cells. If this delivery and removal system stops working, our cells can't maintain homeostasis.

Our blood vessels are constantly involved with regulating our blood flow. The sympathetic division of the nervous system is constantly controlling the smooth muscle found around blood vessels. If this muscle is contracted blood flow to an organ will decrease, if this muscle is relaxed blood flow will increase. You have probably seen this change in blood flow during exercise. If you have started running and felt a pain in your side, this pain was caused by a change in blood flow. When you started to run, more blood was sent to organs needed for physical activity, like the skeletal muscles, heart and lungs. As more blood was sent to these organs, less was sent to others. As blood flow was reduced to the GI tract, the small intestine didn't receive enough oxygen. When an organ becomes oxygen deprived, it will hurt. Think about how someone's heart will hurt during a heart attack.

Our blood vessels are separated out into two main categories, arteries and veins. An artery always carries blood away from the heart. Think of the "a" in artery as standing for away. Veins always return blood to the heart.

There are several generalizations we can make about arteries and veins. These generalizations don't apply to all of them, but we do see the following.

Arteries
1. Arteries carry blood away from the heart.
2. Most arteries will be high in oxygen content.
3. Arteries have thick walls.
4. Arteries are high in pressure.
5. Arteries don't contain valves.
6. Arteries are deep within the body.

Veins
1. Veins carry blood to the heart.
2. Most veins will be low in oxygen content.
3. Veins will have thin walls.
4. Veins are lower in pressure.
5. Veins contain valves.
6. Many veins are superficial.

Our arteries and veins change in structure as they travel away from and back to the heart. The different types of blood vessels are:

1. Elastic arteries. The first arteries will be found at the base of the heart and begin with the aorta. These large elastic arteries get their name because of the large amount of elastic fibers we find in them. These arteries need this elastic tissue so they will stretch when blood pressure rises. We have all heard of "hardening of the arteries" and what happens when they harden? Our blood pressure rises. This elastic nature of arteries helps to stabilize our blood pressure and ease blood flow.

The large elastic arteries are primarily were we see age related changes, leading to disease. Two big problems we see over time are atherosclerosis and arteriosclerosis. Atherosclerosis refers to the deposition of fats (plaque) in the artery wall. This leads to an artery being filled with materials, which narrows the center (lumen) of the blood vessel. As the artery narrows, less blood is transported. As less blood is transported, less oxygen is delivered and tissues may start to die. Arteriosclerosis refers to the loss of elasticity over time. As arteries become less elastic, they won't stretch. When they won't stretch we are left with smaller pipes to push blood through. Smaller pipes mean less blood and oxygen delivery.

2. Muscular arteries. As we move away from the heart, our blood pressure drops and less difference is seen between the systolic and diastolic pressure. As this pressure decreases, less elastic tissue is needed. As the arteries lose the elastic tissue, they gradually contain more smooth muscle. This smooth muscle is used to regulate blood flow. This is where much of the vasoconstriction and vasodilation occurs. The muscular arteries are also called distributing arteries.

3. Arterioles. Arterioles are the smallest of arteries and the last arteries seen before a capillary.

4. Metarteriole. These are the short connections between an arteriole and a capillary. Precapillary sphincters will regulate how much blood will move through the capillary or how much the capillary is bypassed by a thoroughfare channel.

5. Capillaries. The capillaries are very important because this is where most all materials move in and out of the blood. Most all movement occurs here, because the capillary wall is a simple squamous epithelial layer called the endothelium. Thin walls make diffusion easier, so most material is moved through the capillaries. This endothelium can be found throughout the cardiovascular system.

Capillaries come in three variations: continuous, fenestrated and sinusoidal.

Continuous capillaries are the most common type of capillary. These cells get their name because the cells don't have any space between them. These capillaries aren't as permeable as other capillaries, so less materials will pass in and out of the blood. Of course lipid soluble materials will still move freely through the plasma membrane.

Fenestrated capillaries get their name because of the tiny gaps between the cells. These gaps make them more permeable and can be found in large numbers in the kidneys. With the tiny holes throughout this capillary, they become more permeable than a continuous capillary. Imagine a soaker hose in some ones garden. A soaker hose is a regular garden hose with many holes in it. When you pressurize the hose, water streams out of all the holes. A fenestrated capillary must look like this, when the plasma leaves the fenestra.

Sinusoidal capillaries have even larger gaps between the squamous cells. With larger gaps, larger materials may move between the cells.

6. Venules. Venules are the first veins seen after the capillaries. The blood is now on its way back to the heart.

7. On the way back to the heart, the remaining veins will be the small, medium and then large veins like the vena cava. The veins will contain valves, which are made of an overlapping of epithelial cells. Like valves everywhere, valves always have one very important function, preventing backflow.

BLOOD VESSEL LAYERS

Blood vessels except for the capillaries will possess three layers, also called tunics. The tunica intima is the deepest layer, so it is the layer which makes contact with the blood. The simple squamous endothelium is found on the inside and is very smooth. Being smooth reduces friction and makes the blood flow easier. The endothelium is surrounded by a basement membrane, just like most epithelial tissues.

The middle layer is the tunica media and is where most of the smooth muscle will be found. This smooth muscle is used to alter blood flow to tissues. Constriction will result is less blood and dilation will allow greater blood flow to a tissue. A noticeable elastic layer is found in the tunica media also.

The tunica externa is the superficial layer of the three. This thick, strong outer layer makes a tough covering around the vessel. Collagen will make this layer strong, just like collagen does in many tissues.

OTHER VESSEL TYPES

Portal systems. A portal system will always be two capillary networks with a direct connection in between. Think of the connection between them as a one way road. With a one way road, materials entering the first capillary network can only go to the second capillary and nowhere else. The body sets up portal systems so that materials can only go from point A to point B and nowhere else.

Anastomosis. An anastomosis is basically a capillary bypass. They involve direct connections allowing blood to avoid a capillary.

Vasa vasorum. These are tiny blood vessels penetrating the wall of a thick walled vessel. Thick walled arteries like the aorta have so many cell layers, that diffusion isn't enough to penetrate to the deeper cell layers. So blood vessels penetrate the thick walls and extend to the deeper cells to provide nutrients and remove wastes.

ARTERIES

You must be familiar with the major arteries of the body. We will start at the base of the heart and work our way out to all major arteries. Remember that arteries carry blood away from the heart, you must consider this when looking at blood flow. Often you will have questions asking which artery leads to the next. The blood will always be traveling away from the heart with an artery question. With a vein question, you must consider the opposite. With veins they will be carrying blood back to the heart. Forgetting the direction of the blood flow is a very common mistake.

Aorta. The first artery leaving the base of the heart is the aorta and remember the base of the heart is the broad superior part. The aorta has several sections to it: ascending, arch, descending, thoracic and abdominal.

Ascending aorta. The first part of the aorta is the ascending aorta, thus named because it is the short piece ascending up in a superior direction from the base of the heart. There are two arteries coming off of the ascending aorta, the two coronary arteries. The coronary arteries are bringing blood back to the cardiac muscle. So notice as the left ventricle pushes blood into the aorta, some of that blood comes back to the cardiac muscle itself. The heart is pumping blood to itself, just as its pumping blood to everything else.

Aortic arch. The next part of the aorta is the arch. The aortic arch gets its name because the aorta makes a 180 degree turn and goes from ascending to descending. This arch has three arteries branching off of it. The three arteries branching off of it are the brachiocephalic, left common carotid and left subclavian.

Brachiocephalic artery. There is only one brachiocephalic artery and it moves off to the right side of the body. It will be supplying blood to the right side of the chest, right side of neck and head and right upper limb.

Left common carotid. The left common carotid is the second branch off of the aortic arch and feeds blood to the left side of the head and neck. There is also a right common carotid, which comes off of the brachiocephalic artery.

Common carotids. The two common carotids travel up the left and right side of the neck. Just before you get to the angle of the mandible, the common carotids split at a thick spot called the carotid sinus. The two branches off the common carotids are the internal and external carotids.

Internal carotids. The two internal carotids (left and right) are the two major inflows of blood to the brain.

External carotids. The external carotids run laterally along the head and supply blood to everything superficial to the brain. The flat bones of the skull, a few muscles and the scalp are supplied blood.

Blood flow to the brain. In addition to the two internal carotids, we also have two vertebral arteries taking blood to the brain. The vertebral arteries travel through the transverse foramen of the cervical vertebrae. Remember the cervical vertebrae have holes in the transverse processes, these holes are for arteries and veins. So the brain has 4 inflows of blood: two internal carotids and two vertebral arteries.

Left subclavian artery. The left subclavian artery is the third branch off of the aortic arch. This artery supplies blood largely to the left upper limb.

Blood flow to the upper limbs. Both upper limbs receive blood from the subclavian arteries and remember blood is moving away from the heart in an artery. If we follow this blood on out, we will see the following sequence of arteries. Subclavian lead to the axillary arteries found under the arm in the armpit area. Axillary leads to brachial arteries found in the brachial region. Brachial artery leads to the radial and ulnar arteries. These two lead to the palmar arteries. The palmar lead to the digital arteries in the fingers.

Subclavian arteries. The arteries passing under the clavicle and continuing towards the upper limb. The shoulder and chest are supplied blood through these arteries.

Axillary arteries. These arteries pass under the arm and supply blood to the shoulder and chest.

Brachial arteries. A large artery extending from the shoulder to the elbow, supplying the upper limb with blood. The brachial artery is commonly used when taking blood pressure.

Radial arteries. The artery found laterally on the thumb side of the forearm, supplying the lateral part of the forearm with blood.

Ulnar arteries. The artery found medially on the little finger side of the forearm, supplying the medial part of the forearm.

Palmar arteries. Arteries found in the palm of the hand.

Digital arteries. Arteries found in the fingers (or toes).

Descending aorta. The descending aorta has two sections to it. The first part is the thoracic aorta and it is the part of the descending aorta found in the thoracic cavity. The second part of the descending aorta is the abdominal aorta. When the descending aorta passes through the diaphragm, it becomes the abdominal aorta.

Thoracic aorta. The thoracic aorta supplies blood to many organs in the thoracic region.

Abdominal aorta. The abdominal aorta supplies blood to the visceral organs of the abdominal cavity. The abdominal aorta ends as it splits into the two common iliacs.

Celiac trunk. Supplies blood to many abdominal organs.

Splenic artery. Supplies blood to spleen and pancreas.

Gastric artery. Supplies blood to stomach.

Common hepatic artery. Supplies blood to liver.

Phrenic artery. Supplies blood to diaphragm.

Superior mesenteric. Supplies blood to small intestine and colon.

Suprarenal arteries. Supplies blood to the adrenal glands.

Renal arteries. Supplies blood to the kidneys.

Gonadal arteries. Supplies blood to ovaries and testis.

Common iliacs. At the second lumbar vertebrae the abdominal aorta will divide into the two common iliacs. The common iliacs continue downward toward the pelvis and lower limbs.

Internal iliacs. The internal iliacs branch off the common

iliacs and supply blood into the pelvis.

External iliacs. The external iliacs continue past the pelvis toward the lower limbs.

Femoral arteries. Large arteries found from the hip to the knee.

Popliteal arteries. Arteries found behind the knees.

Tibial arteries. The tibial arteries follow the tibia down the medial side of the leg. The tibial will divide into the anterior and posterior tibial arteries.

Fibular arteries. This artery travels laterally down the leg.

Dorsalis pedis. Travels down to the foot.

Digital arteries. Found inside the toes.

VEINS

While the body tries to keep all arteries deep, many veins are superficial. The body tries to keep the arteries deep, because of the high amount of pressure inside of them. If an artery is severed, blood loss will be very rapid. Veins don't have nearly as much pressure, so they may run superficial. The name of a vein will tell you if it is superficial. If a vein has the same name as an artery, then the vein runs alongside the artery, meaning it is deep. If a vein has a name different from any artery then the vein is superficial. For example, let's look at the brachial vein versus the cephalic vein. The brachial vein has the same name as the brachial artery, which means the vein runs deep. The cephalic vein doesn't have the same name as any artery, which means this vein is superficial.

If we start at the heart, we have three inflows into the right atrium. The superior vena cava returns blood from above the heart, the inferior vena cava returns blood from below the heart and the coronary sinus returns blood from the cardiac muscle.

Superior vena cava. Returns blood from the two brachiocephalic veins and returns all blood from above the heart.

Brachiocephalic veins. Returns blood from the head, neck and upper limbs.

Subclavian veins. Returns blood from the axillary region and part of the chest.

Axillary veins. Returns blood from the basilic and brachial veins.

Cephalic veins. Returns blood from the lateral region of the upper limb.

Basilic veins. Returns blood from the medial region of the upper limb.

Brachial veins. Returns blood from the deep region of the upper limb.

Median cubital veins. The veins seen on the anterior surface of the elbow and is often used to draw blood from.

Radial veins. Returns blood from the lateral part of the forearm.

Ulnar veins. Returns blood from the medial part of the forearm.

Palmar arches. Returns blood from the hand.

Digital veins. Returns blood from the fingers.

The brain is drained by four veins. The internal jugulars are the two largest returns and the two vertebral veins are the others.

External jugulars. Return blood from the head and neck.

Facial vein. Returns blood from the face.

Coronary sinus. Returns blood from the cardiac muscle.

Azygos vein. Returns blood from the thoracic wall and many thoracic organs.

Phrenic vein. Returns blood from the diaphragm.

Hepatic vein. Returns blood from the liver.

Hepatic portal vein. Returns blood from the GI tract to the liver. A rare example of a vein draining blood into an organ.

Cystic vein. Returns blood from the gallbladder.

Gastric vein. Returns blood from the stomach.

Splenic vein. Returns blood from the spleen.

Inferior mesenteric vein. Returns blood from the colon and rectum.

Gonadal veins. Returns blood from the ovaries and testes.

Superior mesenteric vein. Returns blood from the small intestine and colon.

Inferior vena cave. Returns blood from below the heart.

Common iliac veins. Returns blood from the lower limbs into the inferior vena cava.

External iliac veins. Returns blood from the lower limbs to the common iliac.

Femoral veins. Returns blood from the lower limbs into the external iliac and is found from the knee to the hip.

Great saphenous veins. The longest blood vessel of the body, runs the entire length of the lower limb.

Popliteal veins. Returns blood from the leg region and is found behind the knee.

Tibial veins. Returns blood from the medial region of the leg.

Fibular veins. Returns blood from the lateral region of the leg.

Digital veins. Returns blood from the toes.

BLOOD FLOW

Blood moves through vessels in two ways, this is what's called laminar flow and turbulent flow. Laminar is a smooth, rapid flow seen in large, straight vessels like the aorta. The blood moves in many concentric layers, meaning one layer inside of another. The superficial layers, close to the vessel wall move the slowest and the deeper you go, the faster the blood moves. You can't think of all the blood moving at the same speed. The closer you get to the center, the faster it is moving.

Turbulent flow is a chaotic flow, seen anywhere a vessel turns, narrows or encounters an obstruction. This flow will be much slower than laminar flow. This blood is under more resistance and requires more pressure to move.

These two flows can be related to someone having their blood pressure taken. When a blood pressure cuff (sphygmomanometer) is placed around some ones brachial artery, the valve is closed and the bulb is pumped. As the bulb is pumped, the cuff squeezes the artery until blood flow through the artery is stopped. The valve is then turned to begin letting off the pressure. As the pressure drops, the blood starts to squeeze through the brachial artery and someone with a stethoscope can hear the blood movement. As the blood squeezes through the constricted artery, this sound can be heard. Think about someone pinching a garden hose. When the end of the hose is pinched the water squeezes through and makes a sound as it passes. This is the sound of turbulent blood flow through the artery and this first passing of the blood through the artery, gives us the systolic reading. This reading is usually around 120 mmHg for the average resting adult. Remember that systole refers to contraction. What is contracting at this time? The left ventricle is contracting to push blood through this artery. Pressure is greatest in an artery at this time, so that is when we get our greatest pressure reading on the gauge of the blood pressure cuff.

As we continue to let off the pressure of the cuff, the artery will eventually open fully. When the artery is no longer being pinched, but returns to a wide open artery once again, then the sound of the blood flow changes from turbulent to laminar. When the blood flow changes, the sound of the blood flow changes and this can be heard. Think of when a water hose is being pinched, versus a water hose that is not. We hear two different sounds in each case. When the artery opens back fully and the change in sound is detected, we get our second reading the diastolic. This will tell us the lowest amount of pressure on the artery wall and is usually around 80 mmHg.

So systolic is the greatest amount of pressure in the artery and diastolic is the least amount of pressure. The systolic occurs when the left ventricle contracts and the diastolic occurs when the left ventricle relaxes. This will give us the greatest and least amount of force on the artery wall.

Why do we take someone's blood pressure? We want to know the pressure in the arteries, because this tells us how hard the heart has to work. The higher the blood pressure the more force the cardiac muscle has to generate. The more force it has to generate, the more oxygen it has to consume. If our heart ever gets to a point where it is using more oxygen than can be delivered to it, someone will have a heart attack. If adequate blood flow isn't restored to cardiac muscle within 20 minutes, the muscle will die and will be lost forever.

What causes our blood to flow? The answer is pressure gradients. A pressure gradient means you have two areas of different pressures. Materials will always move from the area of high to low pressure. When the left ventricle contracts, it generates a high pressure area. The blood can only move out the aorta towards a lower pressure area and it does. The further you move away from the heart (the pump) the lower pressure gets. So blood will flow away from the heart and out to all of the body.

The basic blood flow equation is P1-P2/R. Blood flows from the P1 (high pressure) to the P2 (lower pressure), but we must also consider the resistance to the blood flow.

Resistance is caused by many things within the cardiovascular system. Blood must pass through constrictions, turns, obstacles, etc. All of this will slow the blood flow. The resistance equation comes in many forms, but no matter the form, the two important variables will always be viscosity and radius.

Viscosity refers to how thick the blood is. If our blood is unusually thick for any reason, then it will be more difficult to pump. The more difficult it is to pump the more resistance we have. When resistance increases, blood flow decreases. We want the resistance to be as low as possible. The lower the resistance, the less work our heart must do to move it. The relationship between resistance and flow is called Poiseuille's Law.

Our viscosity is usually changed in a few ways. When we increase hematocrit or decrease plasma volume. If our oxygen delivery is inadequate for any reason (like smoking), our body will make more red blood cells, this will result in an increased hematocrit. If we have more red blood cells in our blood, this will increase viscosity. An increased viscosity means more resistance and less blood flow. If we decrease our plasma levels for some reason, like if we have been sweating for hours, we will also increase viscosity.

The most important variable when it comes to resistance is by far radius. Radius refers to the radius of the blood vessel. Our blood vessels have smooth muscle around them, so the nervous system can change the size of the blood vessels when it wishes. By changing the radius only a small amount we get large changes in resistance. When the radius decreases, resistance increases. If our body wants to send less blood to an organ, it will only need to constrict the blood vessels a small amount, because this will give a big increase in resistance.

Laplace's Law states that the total amount of force inside a blood vessel is proportional to the diameter multiplied by the pressure. $F = D \times P$. This law tells us that the larger the diameter of a blood vessel and the more pressure that is in it, the more force we will find against the walls. This information is relevant to aneurysms. An aneurysm is not a ruptured artery, but one with a weak spot creating a bulge. If an artery wall becomes weak for any reason, the pressure tends to push the weak spot outward. As the spot enlarges it weakens more and may eventually rupture. Consider where aneurisms are most likely to occur, where diameter is greatest and where pressure is greatest. This means that the large arteries close to the heart are the most likely to have aneurysms (aorta). If these large, high pressure arteries were to rupture, blood loss would be great. This is why aneurysms often cause death.

CAPILLARY EXCHANGE

Capillary exchange refers to the movement of materials in and out of the capillaries at the tissues. The capillaries are the smallest and thinnest of our blood vessels. With the walls being simple squamous epithelial tissue, this will make movement easier. The thinner the wall the easier it is for diffusion to take place and diffusion is the primary process by which materials move in and out of the capillaries.

On the arterial end of the capillary pressure is high and on the venous end of the capillary the pressure is low. This difference in these pressures facilitates the movement of materials. On the artery side the pressure in the artery is greater than the pressure in the tissue, so this will help move materials out of the blood and into the tissue. As materials leave the blood to enter the tissue, pressure in the capillary drops. By the time we get to the venous end of the capillary pressure has dropped below tissue pressure and materials move back into the blood. This movement out and into the capillaries, along with diffusion and osmosis, will provide the swapping of materials needed to keep the cells alive. Nutrients will be brought into the cells and wastes will be brought out.

When we look at the volume of materials brought into a tissue on the artery side, the venous side takes 90% of that material back out. The remaining 10% is taken out of the tissue through the lymphatic system. The fluid will be called lymph inside the lymphatic vessels and will be cleaned along the way. Eventually this fluid will be returned to the cardiovascular system, where it came from to start.

MONITORING BLOOD PRESSURE

Baroreceptors are one way the brain monitors our blood pressure. A baroreceptor monitors pressure, you can also think of these as stretch receptors. In the large arteries like the aorta and the internal carotids, we have these baroreceptors. The aorta is a good place for this receptor, because this is the first artery of the body. If you know what pressure is in the aorta, then you have an idea of what pressure is everywhere. If pressure was lost in the aorta, pressure would be lost everywhere, so this is a good monitoring site. The internal carotids had better maintain pressure or the brain won't be supplied with enough blood. This is why we see baroreceptors in these two places.

A baroreceptor works by stretching and the brain monitors them by frequency of action potentials. When our blood pressure is low, our arteries are stretched a small amount. When they are stretched a small amount, they send action potentials back to the brain infrequently. The brain sees a low frequency as low blood pressure. When our blood pressure is high, our arteries are stretched more. The more our arteries are stretched, the more frequently our baroreceptors send action potentials back to the brain. The brain sees a high frequency as high blood pressure. If our brain detects our pressure dropping, it will elevate the heart rate and if it detects our pressure rising to high it will lower the heart rate.

Chemoreceptors detect chemicals like oxygen and carbon dioxide. Of the two chemicals carbon dioxide is the major regulating chemical. We often think oxygen is what is always regulating our heart rate and ventilation, this isn't true. Oxygen levels must change a large amount before we see a noticeable change in heart rate and ventilation. If our oxygen levels drop, chemoreceptors will detect this drop and increase our heart rate and ventilation. If heart rate and ventilation increase, this will deliver more oxygen to the tissues.

Carbon dioxide levels only need to change a very small amount before we see changes in heart rate and ventilation. Why is carbon dioxide the major regulator of the chemicals? It is all about pH levels of the body. Our pH needs to be very close to 7.4 and must stay in the range of 7.35-7.45. This is a very narrow range. Remember the chemical reaction:

$$CO_2 + H_2O \leftrightarrow H_2CO_3 \leftrightarrow H + HCO_3$$

This chemical reaction shows us, whatever happens to carbon dioxide, happens to hydrogen ion. So our body is so sensitive to carbon dioxide levels because of the change in pH. If a person has a heart or lung problem, this could easily lead to an increase in carbon dioxide in the blood. This would cause an increase in hydrogen ions and would result in acidosis. We can't tolerate acidosis or alkalosis, because the proteins in our body are very pH specific. They only keep their proper shape when in the normal pH range. When they are taken out of this pH range, the proteins change shape and they can no longer perform their functions. When proteins malfunction, we deviate from homeostasis and death could result.

If carbon dioxide levels increase in the body, we would see an increase in heart rate and respiration. This would expel the excess carbon dioxide, which would balance the hydrogen ion and bring us back to homeostasis.

OTHER BLOOD PRESSURE REGULATORS

Several hormones of the body work to balance blood pressure. Anti-diuretic hormone (ADH), aldosterone and atrial natriuretic hormone all work to balance water in the body. Water balance and blood pressure always go hand and hand. ADH and aldosterone both work on the kidneys to increase water absorption, this decreases urine output. When would our body want to hold water? When our blood pressure is low or the quantity of water in the blood is insufficient. The water lost in urine, was plasma before it was made. So when we release urine, we are removing water from the blood and this drops our blood pressure. When blood pressure is low our body will release ADH and aldosterone to increase water absorption and this will help to raise our blood pressure. This is a simple negative feedback mechanism.

Atrial natriuretic hormone (ANH) works opposite of ADH and aldosterone. This hormone is released from the right atrium of the heart when blood pressure is high. When our blood pressure is elevated, the right atrium is stretched, this stretching causes the release of ANH. When blood pressure is high, we want to release water. So ANH increases urine output, which is a loss of water. As we lose water our blood pressure will drop. This is another example of negative feedback.

Chapter 4 – Study Questions

1. Arteries are defined as blood vessels which?
a. carry blood to the heart.
b. are low in pressure.
c. have thin walls.
d. are found superficially.
e. carry blood away from the heart.

2. Most arteries will?
a. be oxygen rich.
b. be low in pressure.
c. contain valves.
d. be found mostly in the heart.
e. all of the above.

3. What do veins have that arteries don't?
a. more red blood cells.
b. valves.
c. more plasma.
d. less plasma.
e. none of the above.

4. The arteries closest to the heart are the?
a. large elastic arteries.
b. medium muscular arteries.
c. arterioles.
d. capillaries.
e. venules.

5. Vasoconstriction and vasodilation will occur mostly in?
a. large elastic arteries.
b. medium muscular arteries.
c. arterioles.
d. capillaries.
e. venules.

6. The smallest of all arteries are the?
a. large elastic arteries.

b. medium muscular arteries.
c. arterioles.
d. capillaries.
e. venules.

7. Most all materials move in and out of the blood at the?
a. large elastic arteries.
b. medium muscular arteries.
c. arterioles.
d. capillaries.
e. venules.

8. The first veins after a capillary are the?
a. venules
b. small veins
c. medium veins
d. large veins
e. none of the above

9. The most common type of capillary is the?
a. continuous
b. fenestrated
c. sinusoidal
d. open
e. none of the above

10. Some capillaries have tiny gaps in between the cells, this would be which type of capillary?
a. continuous
b. fenestrated
c. sinusoidal
d. open
e. none of the above

11. Having gaps between the capillary cells will make a capillary?
a. more porous
b. less porous

c. no change

12. Blood vessels have a smooth, inner layer called the?
a. venule
b. endothelium
c. valve
d. fenestra
e. sinusoid

13. The superficial layer of a blood vessel is the?
a. tunica interna
b. tunica media
c. tunica externa

14. Vasoconstriction and vasodilation is controlled through which tunic?
a. tunica interna
b. tunica media
c. tunica externa

15. The layer which comes in contact with the blood is which layer?
a. tunica interna
b. tunica media
c. tunica externa

16. The first artery coming out of the heart is the?
a. brachiocephalic artery
b. common carotids
c. aorta
d. coronary arteries
e. left subclavian artery

17. The arteries coming off the ascending aorta are the?
a. brachiocephalic artery
b. common carotids
c. aorta
d. coronary arteries

e. left subclavian artery

18. The first branch off the aortic arch is?
a. brachiocephalic artery
b. common carotids
c. aorta
d. coronary arteries
e. left subclavian artery

19. The second branch off the aortic arch is the?
a. left common carotid
b. right common carotid
c. left subclavian
d. right subclavian
e. internal carotid

20. The two largest inflows of blood to the brain are?
a. common carotids
b. internal carotids
c. external carotids
d. subclavian
e. vertebral

21. The arteries running under the arms?
a. subclavian
b. brachial
c. axillary
d. radial
e. ulnar

22. The arteries found from the shoulder to the elbow are?
a. subclavian
b. brachial
c. axillary
d. radial
e. ulnar

23. The artery lateral in the forearm?

a. subclavian
b. brachial
c. axillary
d. radial
e. ulnar

24. The artery medial in the forearm?
a. subclavian
b. brachial
c. axillary
d. radial
e. ulnar

25. What artery supplies blood to the liver?
a. thoracic
b. hepatic
c. digital
d. renal
e. gonadal

26. Blood is supplied to the ovaries and testes by the?
a. thoracic
b. hepatic
c. digital
d. renal
e. gonadal

27. Blood is supplied to the kidneys by the?
a. thoracic
b. hepatic
c. digital
d. renal
e. gonadal

28. What artery is lateral in the leg?
a. tibial
b. fibular
c. digital

    d. dorsalis pedis
    e. external iliacs

29. Blood from above the heart is returned to the right atrium by the?
    a. superior vena cava
    b. inferior vena cava
    c. coronary sinus
    d. subclavian veins
    e. brachiocephalic veins

30. What veins deliver blood to the superior vena cava?
    a. axillary
    b. inferior vena cava
    c. coronary sinus
    d. subclavian veins
    e. brachiocephalic veins

31. Blood from the cardiac muscle is returned to the right atrium by the?
    a. superior vena cava
    b. inferior vena cava
    c. coronary sinus
    d. subclavian veins
    e. brachiocephalic veins

32. Blood is commonly taken from the?
    a. ulnar vein
    b. radial vein
    c. brachial vein
    d. median cubital vein
    e. basilic veins

33. The two largest outflows of blood from the brain?
    a. facial vein
    b. internal jugular
    c. external jugular
    d. vertebral

e. azygos

34. Blood is returned from the diaphragm by the?
a. cystic
b. cephalic
c. phrenic
d. gastric
e. gonadal

35. The longest blood vessel in the body is the?
a. femoral
b. tibial
c. fibular
d. great saphenous
e. digital

36. A long, straight vessel will have what type of flow through it?
a. laminar
b. turbulent

37. Blood pressure is usually taken on which artery?
a. femoral
b. tibial
c. brachial
d. aorta
e. popliteal

38. A person would take a systolic pressure reading when hearing which blood flow?
a. laminar
b. turbulent

39. What causes blood to flow?
a. concentration gradients
b. pressure gradients
c. osmotic gradients
d. sodium gradients

e. all of the above

40. The most important variable in determining resistance is?
a. pressure
b. viscosity of blood
c. radius of blood vessel
d. oxygen content
e. number of white blood cells

41. Laplace's law helps us to determine where we are most likely to find?
a. red blood cells
b. clots
c. aneurysms
d. capillaries
e. all of the above

42. A baroreceptor detects changes in?
a. pressure
b. chemicals
c. temperature
d. tension
e. sound

43. Chemoreceptors are most sensitive to?
a. oxygen
b. nitrogen
c. carbon dioxide
d. bicarbonate
e. calcium

44. Normal pH for most of the body is?
a. 7.3
b. 7.4
c. 7.5
d. 8.0
e. 8.2

45. Arteriosclerosis is
a. the low resistance to blood flow.
b. the pressure in the pulmonary arteries.
c. the loss of the tunica media.
d. the gaining of red blood cells.
e. the loss of elasticity in vessels.

46. A clot free floating in the blood is?
a. an embolism
b. a thrombus
c. arteriosclerosis
d. atherosclerosis
e. hemorrhage

47. Which vein would be superficial?
a. brachial
b. femoral
c. radial
d. digital
e. cephalic

48. If blood pressure is low, you would expect the body to respond by doing what with ADH?
a. releasing more
b. releasing less

Answers to multiple choice questions.
1. E
2. A
3. B
4. A
5. B
6. C
7. D
8. A
9. A
10. B
11. A
12. B
13. C
14. B
15. A
16. C
17. D
18. A
19. A
20. B
21. C
22. B
23. D
24. E
25. B
26. E
27. D
28. B
29. A
30. E
31. C
32. D
33. B
34. C
35. D
36. A
37. C

38. B
39. B
40. C
41. C
42. A
43. C
44. B
45. E
46. A
47. E
48. A

LYMPHATIC SYSTEM

The Lymphatic System

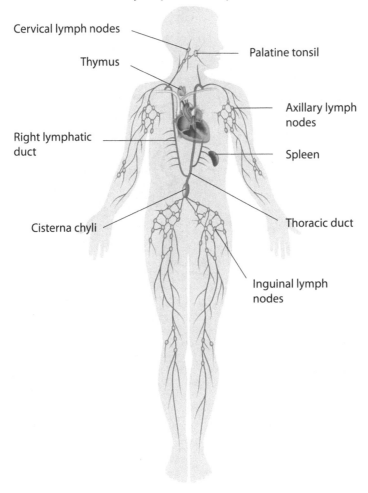

Anatomy of a Lymph Node

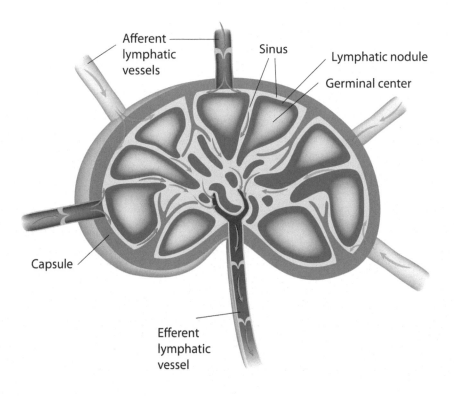

CHAPTER 5
LYMPHATIC SYSTEM

The lymphatic system combined with the cardiovascular system makes the circulatory system. Our circulation is not entirely within the cardiovascular system, our lymphatic system picks up and returns 10% of the fluids from the tissues. Our arteries will deliver all of the materials to the tissues, but the veins only return 90% of what the arteries brought in. So fluid balance is just one of the important functions of the lymphatic system. If lymphatic vessels become clogged, which can happen with some diseases, then fluid will quickly accumulate in tissues. This fluid accumulation will be seen as swelling occurs. Sometimes when a cancer is removed, the lymphatic vessels and nodes draining the infected area will be removed. After removal the tissues often swell until the circulatory system adapts.

We have special lymphatic vessels along the small intestine involved with fat absorption. These lymphatic vessels in this location are called lacteals. Without this part of the lymphatic system we wouldn't be able to absorb adequate amounts of lipids.

The most well-known function of the lymphatic system is fighting disease. We wouldn't live long without a functioning immune system. Think about how often children are sick, it takes many years to fully develop the lymphatic system. We will look at the different organs of this system and see how they protect us from harmful chemicals, foreign invaders and disease from within our own body.

FLUID BALANCE

We have looked at capillaries in the previous chapter and we will see a slightly different type of capillary in this chapter. Lymphatic capillaries have a structure very similar to continuous capillaries, with a few exceptions. The capillaries of the cardiovascular system are continuous with all of the arteries, veins and heart, this is not so with lymphatic capillaries. Lymphatic capillaries begin as dead end tubes in most tissues, they are not one big circular flow of fluid, like we see with blood. These capillaries still have simple squamous epithelial cell layers, but the cells loosely overlap. With this type of loose connection, fluids can enter this capillary type easier, than what the fluid could with a continuous capillary. In addition, the basement membrane is lacking on this epithelial tissue. Without the surrounding basement membrane and the loose overlapping of the cells, these capillaries can pick up tissue fluids easily.

As soon as interstitial fluid enters a lymphatic capillary, the fluid is then called lymph. We have names for many fluids around the body and we see a few with the circulatory system. The fluid material within the blood is called plasma. When this fluid leaves a capillary and enters a tissue, the fluid is then called interstitial fluid. When that fluid leaves a tissue and enters a lymphatic capillary, it is then called lymph. The lymphatic vessels will return the lymph fluid back to the cardiovascular system through the subclavian veins. All of this fluid came from the blood initially, so it is returned there. Materials are being swapped, but the fluid is all very similar.

Lymphatic vessels along the small intestine are doing the same thing as others, but they will gather lipids as the GI tract absorbs them. These special vessels called lacteals follow the small intestine, because the small intestine is where we absorb almost everything we eat or drink. As we eat fats the small intestine will absorb them, the lacteals will pick them up and return all of the material back to the cardiovascular system. This lymph containing fats is called chyle.

The lymphatic vessels are very similar to veins in structure. Like veins lymph vessels have valves. Like valves anywhere, they are preventing the back flow of fluids. The tiny lymph capillaries will lead to lymph vessels and eventually back to the large ducts. The thoracic duct accepts lymph drainage from the entire lower half of the body, plus the left side of the head, neck, chest and abdomen. The right lymphatic duct accepts lymph from the right side of the head, neck, chest, abdomen and right upper limb.

Since lymphatic vessels are not part of the cardiovascular system, the fluids are not being pumped by the heart. The fluid is returned by several mechanisms. As fluid and materials leave the blood and enter a tissue, this will create a small amount of pressure and help to move materials into the lymphatic system. The movement of air in and out of the lungs, will also assist lymph movement. When we contract the muscles of ventilation, we create a negative pressure inside the thoracic cavity. This negative pressure draws in air, and at the same time draws fluids into the thoracic cavity also. So during inspiration, we are helping to draw lymph back towards the heart. Skeletal muscles also assist in lymph movement. When skeletal muscles contract, they squeeze lymphatic vessels, just like they do veins. This squeezing will compress the fluid in the lymphatic vessels, valves will prevent back flow and the lymph gets squeezed back towards the heart. Smooth muscle around the lymph vessels can do the same. All of these mechanisms working together, return the lymph fluid.

LYMPH NODES

As the lymph is returned along the lymphatic vessels, eventually the lymph will encounter a lymph node. Lymph nodes are the sites where lymph is filtered, meaning where the lymph is cleaned. The lymph can pick up many unwanted materials, as it returns from a tissue.

A lymph node is a round structure with afferent vessels bringing fluid into a node and efferent vessels taking the fluid out. In between these two vessels the node will remove unwanted materials. Lymph nodes filter lymph and not blood. Filtering blood is the spleen's function.

A lymph node is surrounded by a skin like layer called the capsule. A capsule can be seen on other structures, so it is not unique to a node. When we look to the inside of a lymph node, we see two major regions; the cortex and medulla. A cortex will always be an outer region of a structure and a medulla will always be a deeper inner region. As lymph flows in through an afferent vessel, the fluid it will flow inside and around the inner part of the capsule. The capsule has extensions called trabeculae leading deeper into the node. As the lymph encounters the trabeculae, it will flow down and along them. This path will take the lymph deeper into the node. As the fluid travels through and deeper into the node, the lymph will encounter many lymphocytes and macrophages. These white blood cells will remove unwanted materials from the lymph.

In the outer cortex collections of lymphocytes called lymphatic nodules will be stimulated to engulf by phagocytosis any unwanted materials from the lymph. In the center of these nodules we find germinal centers, where the lymphocytes will proliferate when detecting the presence of foreign invaders. Think about how a physician may squeeze the area under the mandible, when we have an infection in the pharynx (throat). The lymph nodes in this area, clean the fluid returning from the pharynx. If a foreign invader is detected these white blood cells will rapidly multiply and this will make the lymph nodes swell. We know if the lymph nodes are swollen, there is probably an infected tissue close to the nodes.

These white blood cells of the lymph node are always in a current of lymph, as this fluid flows through the lymph node. Reticular fibers, which are just a fine collagen fiber, hold all of these cells in place. In addition to holding the WBC in place they can also catch other cells, like cancer cells. This is why lymph nodes downstream from a cancerous tissue are often removed. The lymph node acts as a simple mechanical filter. Just as the filter in a coffee pot holds back large particles and allows water to pass, the lymph node holds back cells and allows the lymph to pass. You can think of these fibers as like a spider web and the cells are the spiders. Spiders wait for an insect to be caught by their web and then they will eat it. These WBC in a lymph node are waiting for the reticular fibers to catch materials and then the cells will eat them.

The lymphatic system shows a large amount of variation among the population. Lymphatic vessels, the position of lymph nodes and number of lymph nodes can vary greatly. Most people will have 400-500 lymph nodes and they can be found in groups. They will be found in the cervical region, draining the head and neck. Some are collected in the axillary region, draining the fluid from the upper limbs. A large number are found in the thoracic cavity because of all the foreign materials entering the lungs. About half of all lymph nodes are in the abdominopelvic cavity because of the GI tract. The remaining lymph nodes are in the groin, draining the lower limbs.

TONSILS

Our tonsils are lymphatic tissue seen to the rear of the oral cavity. We have three sets of tonsils and they are the lingual, pharyngeal and palatine. The lingual tonsils are found on the posterior region of the tongue. The pharyngeal are located posterior and superior to the soft palate. The soft palate and uvula are what you see hanging down in the back of your throat. The pharyngeal are also called the adenoids and sometimes need to be removed, if they interfere with breathing. The palatine tonsils are the ones we can see in the rear of the oral cavity. When we have an infection in our pharynx, we can see the pus (accumulation of dead white blood cells) on the palatine tonsils. This pus will accumulate into round structures and are sometimes called tonsil stones.

The tonsils are called nonencapsulated lymphatic tissue, also called mucosa associated lymphatic tissue. Lymph nodes have a skin like structure around it called a capsule. This capsule separates the lymph node from everything around it. We don't want the tonsils separated from the surrounding oral cavity, so tonsils don't have a capsule. We want the lymphatic tissue as exposed as possible to the oral cavity. With this exposure the tonsils can easily detect and respond to the entry of foreign invaders. This type of lymphatic tissue (nonencapsulated) can be found in the mucous membranes of the body. The mucous membranes are found in the four body systems, which open to the outside of the body: lymphatic, respiratory, digestive and reproductive. Foreign invaders can't cause disease, if they can't get in the body. The tonsils are our first line of defense in the oral cavity and they will stop many potential pathogens.

THYMUS GLAND

The thymus gland is just anterior and superior to our heart inside the thoracic cavity, also inside the mediastinum. This gland is considered part of the endocrine system, but is much more a part of the lymphatic system. This organ is largest around the time we reach maturity and slowly decreases in size as we get older. The gland is composed of epithelial cells, just like most endocrine glands are. Within the gland there are two regions, an outer region called a cortex and a deeper region called a medulla.

The thymus gland releases a group of hormones called thymosin, which is responsible for a maturing of the lymphatic system. What is vitally important is that T-cells mature in the thymus gland. Lymphocytes can be divided into two major groups: B-cells and T-cells. The T-cells will undergo a process of positive and negative selection in the thymus gland. The thymus gland will positively select (keep) T-cells which will be helpful and it will negatively select (eliminate) T-cells which would be harmful. This process should prevent T-cells from damaging healthy cells. T-cells are not fully mature until leaving the thymus.

SPLEEN

The spleen is located in the upper left quadrant of the abdomen. The left kidney is on its inferior surface and the stomach meets it medially. The spleen filters the blood in a similar way in which the lymph nodes filter the lymph.

The spleen is not just involved with filtering the blood. The spleen will remove and destroy many of the old, dead red blood cells of the body. We lose millions of red blood cells every second and a large amount of waste is produced by their destruction. Much of this material is cleaned from the blood in the spleen.

The spleen also acts as a reservoir for blood. When we aren't physically active our spleen is larger, because it is filled with blood. As we become physically active, our spleen will decrease in size as the sympathetic nervous system, diverts blood to other organs. Since the spleen contains large amounts of blood, a ruptured spleen can result in rapid blood loss.

The spleen has a capsule surrounding it. Like other capsules, this layer will separate it from other tissues. The capsule has extensions called trabeculae extending deep into the spleen and these trabeculae will separate the spleen into compartments. On the surface of the spleen a hilum is present. A hilum is an area where structures enter and leave an organ. The splenic artery and vein enter and leave the spleen in this area.

Inside the spleen two areas can be seen. These two areas are white pulp and red pulp. The white pulp is filled with white blood cells (lymphocytes) and surrounds the arterial blood supply entering the spleen. White pulp gets its name because it is filled with white blood cells. The white pulp removes unwanted material from the blood as it passes into the spleen. As blood passes deeper into the spleen we begin to see fewer white blood cells and more red blood cells. This where we have red pulp in the spleen. Red pulp is simply the areas where there are fewer white blood cells and more red blood cells.

The lymphocytes and macrophages of the spleen will detect and respond to foreign materials and invaders in the blood. These protective cells will remove any materials needed from the blood. The spleen cleans the blood, like the lymph nodes clean the lymph.

IMMUNITY

Immunity is the ability to resist and destroy foreign chemicals and invaders.

Immunity comes in two forms: innate and adaptive.

INNATE IMMUNITY

Innate immunity includes the parts of the immune system which don't improve over time. This form of immunity is also called nonspecific, because it doesn't respond to one type of invader one way and another invader a different way. Innate immunity always responds to invaders in the same way every time. These are the general defenses of the body and easier to understand. This immunity doesn't improve over time.

Innate immunity works faster than adaptive immunity, but it lacks the ability to remember how to destroy invaders, after it has done it once before. Each type of immunity has its advantages and disadvantages.

The structures in innate immunity include: skin, mucous membranes, nonspecific chemicals, and many cells. These structures can be broken down into different categories. The categories are: mechanical mechanisms, chemicals and cells.

1. Mechanical mechanisms of innate immunity. The mechanical barriers of the body protect us against the entry of foreign materials and invaders. Our skin is a tough outer layer which can prevent the entry of most materials into the body. Without this barrier, we wouldn't live long.

We also have many barriers deeper in the body. The four systems which open to the outside of the body, all have mucous membranes lining them. The mucous membranes are also mechanical barriers.

Tears are always washing the surface of our eyes. Without tears and blinking, we couldn't keep our eyes clean. Invaders would thrive in the warm, wet, nutrient rich environment.

Saliva continuously washes bacteria out of the oral cavity. The bacteria are swallowed and stomach acids will kill most anything.

Cilia in the air passageways sweep out foreign particles, as mucous traps them.

Hydrochloric acid in the stomach will kill most bacteria we swallow.

All of these mechanisms are an important part of immunity, but remember, they don't improve with time. Immune mechanisms which don't improve with time are all part of innate immunity.

2. Chemicals of innate immunity. There are many chemicals involved with immunity and many of them have overlapping functions, while others are unique.

Histamines, kinins, complements, prostaglandins and leukotrines work in a similar way. This chemicals largely work to promote inflammation. Inflammation occurs whenever tissues are damaged. Chemicals cause inflammation by vasodilation and making blood vessels more permeable. If a blood vessels dilates and the blood vessels become more permeable, more materials move from the blood into the tissue. When blood moves out into a tissue, several changes are seen in the tissue. Five symptoms are always seen with inflammation: redness, heat, pain, swelling and disturbance of function. The area becomes red because of all the red blood cells moving into the area. The area becomes hot because our blood moves heat around our body and as more blood comes in, more heat comes in. The area hurts as pain receptors are activated. The area swells as the tissue fills with blood products. Lastly we don't use body parts when they are painful. This is what is meant by disturbance of function.

Complements will also work to destroy bacteria by binding to the surface of the cell and forming a hole in the cell membrane. This will cause a lysing of the cell, resulting in its destruction.

These chemicals will also attract white blood cells by chemotaxis. White blood cells will follow these chemical trails, like a dog following a trail to a rabbit. The white blood cells follow the trails, because where the chemical are being released, the white blood cells are needed.

Pyrogens work by resetting our internal body temperature. The hypothalamus is where we have the thermostat of our body, so this is where the pyrogens work. Fever is what we call the work of the pyrogens on our body. It's thought we have fever to speed up the immunological response and to create an environment to warm for a foreign invader. When fever heats up our body, why do we feel cold? As our internal temperature increases, the environment around us feels cooler. Also, we often sweat when we run a fever and this will make us feel cool.

Interferons get their name by interfering with viral replication. Once a cell becomes infected with a virus, it becomes a virus factory. The body can't save this cell, but it can protect the neighboring cells. An infected cell will produce interferons and these chemical will cause nearby cells to produce antiviral proteins. These antiviral proteins will protect the cells from further infection.

3. Cells of innate immunity. All white blood cells except lymphocytes are involved with innate immunity. Neutrophils, eosinophils, basophils, monocytes, macrophages, etc. are all in innate immunity, because they don't improve over time. This doesn't make these cells less important. We must have them to survive, but these cells don't get better at destroying foreign chemicals and invaders as we get older.

These white blood cells will move to damaged areas, leave the blood and enter the tissues (diapedesis), then follow the chemical trails to invaders (chemotaxis).

Many monocytes will leave the blood and enter the tissues. After entering the tissues, the monocyte will enlarge, at this time it becomes a macrophage. The macrophages will wander the tissues and destroy materials we don't want in the body. A macrophage doesn't die after one phagocytic event, it will engulf many foreign materials and live longer. This is why macrophages are associated with chronic infections.

Natural killer cells are a class of lymphocyte, which act on classes of cells instead of different species. These are the only lymphocytes found in innate immunity.

ADAPTIVE IMMUNITY

Adaptive immunity includes the parts of the immune system, which do improve over time. This form of immunity is also called specific immunity. The lymphocytes are what make adaptive immunity, because they can recognize specific foreign invaders and can kill them faster, after the initial exposure.

Adaptive memory involves memory and specificity. Meaning they recognize and remember specific species of foreign invaders, not just groups.

Adaptive immunity requires antigens for activation. Antigens are any substance capable of stimulating an immune response, causing the production of antibodies. Antigens come in two forms: foreign and self. Foreign antigens are antigens originating outside of the body and this is where most antigens come from. Examples include: bacteria, viruses, foods, drugs, pollen and many other materials. It makes sense that most substances which stimulate an immune response would originate from outside the body. Self-antigens are antigens originating inside the body. We don't have many self-antigens but we do have those originating from cancers. Cancers are cells originating from within our body and most all of the time our immune system destroys them.

Adaptive immunity is divided into two categories, humoral immunity and cell mediated immunity. These two types of immunity exist, because we have two major types of lymphocytes.

Humoral immunity is also called antibody mediated immunity and is associated with the B lymphocytes. Humors refer to fluids of the body and the adaptive immunity in our humors are antibodies.

Lymphocytes are divided into two main groups, T-cells and B-cells. The B-cells are divided into many groups themselves, but many of them develop into B plasma cells. B plasma cells produce antibodies, when stimulated by antigens. These antibodies will destroy a pathogen and direct other cells to attack them. So humoral immunity is all about B-cells and antibodies. Antibodies will be most effective against extracellular invaders.

Cell mediated immunity is associated with T lymphocytes. T cells attack foreign invaders directly, instead of producing damaging chemicals, like a B-cell does. So cell mediated immunity is all about T-cells. T-cells will be most effective against intracellular invaders like viruses and parasites.

In order for lymphocytes to work, they must be activated by an antigen. Activation can occur in two ways: through antigenic receptors or major histocompatibility complexes (MHC). MHCs come in two forms, class I and class II. Class I MHCs are found on the surface of an infected cell and act like a sign, telling nearby WBC to destroy it. Class II MHCs are displayed by white blood cells and act as an alarm to neighboring cells.

Upon activation B-cells will rapidly divide. Many B-cells are made and some develop into B-plasma cells while others develop into B-memory cells. The plasma cells will produce antibodies in huge numbers. The remaining cells don't produce antibodies but become the memory cells. Memory cells will remain and if exposed to the same antigen a second time, they will start antibody production much faster and in much larger quantities. On the first exposure to an antigen, the body requires anywhere from a few days to a few weeks to produce antibodies. On the second exposure memory cells will give antibody production in hours and in much greater numbers.

The first time we are exposed to a foreign invader, it takes days for our immune system to respond. In this time we feel all the symptoms of the disease and move away from homeostasis. On the second exposure our immune system responds much faster and destroys the invader before we ever know it's there. That is immunity, when we can eliminate a foreign invader, before showing any symptoms of the disease.

Antibodies come in several classes. The classes of antibodies, also called immunoglobulins, are identified with letters. IgG (immunoglobulin G) is the most common of the antibodies. These antibodies are associated with the second antibody response to an antigen and can cross the placental barrier. IgM is the second most common antibody and it is associated with the primary response of an antigen. These antibodies are also associated with transfusion reactions of the blood. IgA is associated with mucous membranes, saliva and tears. IgE is associated with allergic responses. IgD functions are not understood.

Antibodies work in several ways. Antibodies can cause inflammation, promote phagocytosis by white blood cells, cause cells to release chemicals and bind to antigens.

Chapter 5 – Study Questions

1. Lymphatic vessels return what % of fluid from tissues?
a. 1%
b. 2%
c. 5%
d. 10%
e. 50%

2. Which is not a function of the lymphatic system?
a. fluid balance
b. fat absorption
c. destroying foreign invaders
d. carrying fluids to tissues
e. none of the above

3. Lymph vessels are most similar to?
a. arteries
b. capillaries
c. arterioles
d. the aorta
e. veins

4. Special lymphatic vessels found along the intestine are called?
a. lacteals
b. lymphatics
c. lymph nodes
d. chyle
e. none of the above

5. A lymphatic capillary is made of what cells?
a. simple cuboidal
b. stratified squamous
c. simple squamous
d. transitional
e. none of the above

6. Fluid inside the lymphatic system is called?

a. interstitial fluid
b. lymph
c. plasma
d. humors
e. synovial fluid

7. Lymph is returned to the cardiovascular system through what veins?
a. inferior vena cava
b. superior vena cava
c. subclavian
d. brachial
e. brachiocephalic

8. Lymph fluid with fats added is called?
a. lacteals
b. lymphatics
c. lymph nodes
d. chyle
e. none of the above

9. The return of lymph from the tissues is assisted by?
a. skeletal muscle contraction
b. ventilation
c. smooth muscle in lymph vessels
d. none of the above
e. all of the above

10. Lymph is filtered by _____, while blood is filtered by _____?
a. spleen, lymph nodes
b. lymph nodes, spleen
c. lymph nodes, thymus gland
d. thymus gland, spleen
e. thymus gland, tonsils

11. A lymph node is surrounded by a layer called the?
a. capsule

b. hilum
c. thymus
d. endothelium
e. malt

12. The outer region of a lymph node is the?
a. medulla
b. cortex
c. hilum
d. thoracic duct
e. lymph layer

13. Which is not part of the lymphatic system?
a. spleen
b. tonsils
c. thymus
d. lymph node
e. kidneys

14. Swollen lymph nodes indicate the presence of?
a. histamines
b. infection
c. interferons
d. complements
e. none of the above

15. Cancers can be trapped in a lymph node by?
a. elastic fibers
b. lymph
c. reticular fibers
d. chyle
e. valves

16. Lymph enters a lymph node through a?
a. artery
b. vein
c. afferent vessel
d. efferent vessel

e. duct

17. Which of the following is true about the tonsils?
a. We have 4 sets of tonsils.
b. The lingual tonsils are located in the pharynx.
c. The tonsils help trap chyle.
d. The pharyngeal tonsils are also called the adenoids.
e. The tonsils filter blood.

18. Enlarged pharyngeal tonsils can interfere with?
a. breathing
b. blood flow
c. action potentials
d. lymph flow
e. chyle delivery

19. T cells mature where?
a. liver
b. tonsils
c. lacteals
d. red bone marrow
e. thymus gland

20. Which is not a function of the spleen?
a. blood reservoir
b. filtering blood
c. destroying old red blood cells
d. fat absorption
e. responding to foreign substances

21. When the sympathetic nervous system is activated the spleen will?
a. increase in size
b. decrease in size
c. no change

22. With innate immunity?
a. the response is always the same.

b. the response improves over time.
c. each exposure is specific.
d. all of the above
e. none of the above

23. Which is found in innate immunity?
a. antibodies
b. humoral immunity
c. cell mediated immunity
d. mucous membranes
e. none of the above

24. White blood cells are attracted to chemicals released by foreign invaders. This attraction is called?
a. hemostasis
b. chemotaxis
c. interferon
d. antigens
e. histamines

25. Which chemicals work best against viruses?
a. interferons
b. antigens
c. histamines
d. complements
e. pyrogens

26. Which chemical stimulate fever production?
a. interferons
b. antigens
c. histamines
d. complements
e. pyrogens

27. Which is not a cell of innate immunity?
a. neutrophil
b. basophil
c. lymphocyte

d. eosinophil
e. monocyte

28. Antibodies are produced by?
a. T cells
b. B cells
c. macrophages
d. neutrophils
e. monocytes

29. Cell mediated immunity is associated with?
a. T cells
b. B cells
c. macrophages
d. neutrophils
e. monocytes

30. A class I MHC is displayed by?
a. a virus infected cell
b. B cells
c. macrophages
d. monocytes
e. dendritic cells

31. Which is not a sign of inflammation?
a. redness
b. heat
c. bleeding
d. pain
e. swelling

32. Adaptive immunity is stimulated by?
a. interferons
b. antigens
c. histamines
d. complements
e. pyrogens

33. The most common antibody is?
a. IgG
b. IgM
c. IgA
d. IgD
e. IgE

Answers to multiple choice questions.
1. D
2. D
3. E
4. A
5. C
6. B
7. C
8. D
9. E
10. B
11. A
12. B
13. E
14. B
15. C
16. C
17. D
18. A
19. E
20. D
21. B
22. A
23. D
24. B
25. A
26. E
27. C
28. B
29. A
30. A
31. C
32. B
33. A

RESPIRATORY SYSTEM

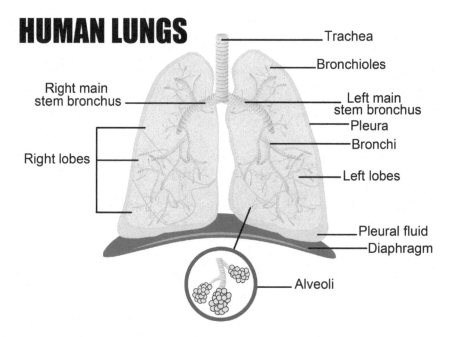

CHAPTER 6
RESPIRATORY SYSTEM

The respiratory system begins at the openings to the nasal cavity and ends in the deepest spaces of our lungs. Even though we can take air in through the oral cavity, the oral cavity isn't part of this system. We will see the nose, pharynx (throat), larynx, trachea, bronchi and lungs.

FUNCTIONS OF RESPIRATORY SYSTEM

The primary function of the respiratory system is to swap oxygen and carbon dioxide within the lungs. We must bring oxygen in so our cells can go through aerobic respiration. Aerobic respiration is the chemical process by which living cells obtain most ATP. We must remove carbon dioxide or the body pH will quickly become acidic.

We need to distinguish between ventilation and respiration. These terms are often used as synonyms, but they aren't the same. Ventilation is the moving of air, in and out of our lungs. Ventilation is accomplished by the muscles of ventilation. We will cover these muscles soon. Respiration is the exchanging of gases and this occurs by simple diffusion. Just because someone has adequate ventilation, it doesn't mean they have adequate respiration. These two processes are independent of each other.

Respiration comes in two forms, internal and external. External respiration occurs at the lungs. During external respiration, oxygen enters the blood and carbon dioxide leaves the blood. This is the gas movement we want at the lungs. Internal respiration occurs at the tissues. At the tissues oxygen leaves the blood and carbon dioxide enters the blood. Notice the same gases are being swapped, but in different directions.

Regulation of blood pH is vital to homeostasis, and our respiratory system is the primary regulator of pH. Don't forget that carbon dioxide and hydrogen ion balance always go hand and hand.

$$CO_2 + H_2O \leftrightarrow H_2CO_3 \leftrightarrow H + HCO_3$$

The chemical reaction shows us that carbon dioxide chemically combines with water to make carbonic acid. The enzyme carbonic anhydrase speeds up the reaction between carbon dioxide and water. Carbonic acid dissociates into hydrogen ion and bicarbonate ion. This reaction tells us two very important things.

1. This is a reversible reaction, which means whatever happens to the materials on one side; will happen to the materials on the other side. All of these materials are trying to reach a balance called equilibrium.

2. Carbon dioxide and hydrogen ion are on opposite sides, so what happens to one will happen to the other. If carbon dioxide levels rise, then hydrogen ion levels will rise. If carbon dioxide levels drop, then hydrogen ion levels will drop. An equilibrium is always being reached between these two materials. This reaction shows us that as long as the respiratory system is balancing carbon dioxide, we will be balancing hydrogen ions. So we can obtain pH balance by balancing carbon dioxide.

When we look at blood pH, we must keep it in the range of 7.35-7.45. If the number gets to low, a condition called acidosis will be the result. Acidosis is when we have too many hydrogen ions in the blood. If the number gets too high, a condition called alkalosis will be the result. Alkalosis is when we have too few hydrogen ions in the blood.

We can't allow pH to deviate in the body, because proteins are very pH specific. If pH moves outside of the 7.35-7.45 range, proteins will change shape. When they change shape, they can no longer do their job. When this happens the body will leave homeostasis. Our proteins are also temperature specific and that is why we must maintain a body temperature close to 98.6 F.

Our sense of smell is associated with the respiratory system. Smell begins high in the superior region of the nasal cavity. We must bring air in to get chemicals to these sensory receptors. So when we inhale through the nose, chemicals are brought into the nasal cavity and allow our sense of smell to work.

We can't speak without air moving past the vocal cords, located inside the larynx. As air moves over the vocal cords, the vocal cords vibrate and these vibrations are primarily responsible for the sound of our voice.

The respiratory system is divided into the upper and lower respiratory tracts. The upper respiratory tract includes the nasal cavity and pharynx (throat). The lower respiratory tract includes the larynx, trachea, bronchi and lungs. Some texts will place the larynx in the upper respiratory tract.

NOSE AND NASAL CAVITY

Our respiratory system starts at the naris, the openings to the nasal cavity. Just inside the naris is a small chamber called the vestibule. This is the space inside of where the nostrils flare. Deep to the vestibule are three passageways called meatus. Air flows through each meatus, in between the three bony ridges called concha. In between the three concha is the nasal septum, which separates the nasal cavity into a left and right half. The nasal septum and concha help to create a very turbulent air flow through the nose. We want air to swirl around in all directions, as it comes through the nose. As the air swirls, particles in the air will hit the mucous and stick to it. This gives a cleaning of the air. If we spend a day on a dusty road and then blow our nose later, we can see how particles are trapped in the nose. In addition the air will be warmed and humidified as it passes through the nose. Our sense of smell is also located in the nasal cavity.

Several bones surrounding the nose have hollow spaces which act as resonating chambers. These chambers affect the sound of our voice. If these chambers fill with mucous, like when we have a cold, the sound of our voice changes. These bones include the frontal, ethmoid, sphenoid and maxilla. They may also be called the paranasal sinuses.

Summary of nasal cavity functions: cleans air, warms air, humidifies air, sense of smell, and affects the sound of the voice.

PHARYNX

As we travel deep to the nasal cavity we move past the auditory tubes (Eustachian tubes). These passageways connect to the middle ear. Usually these passageways are closed, but if we swallow or yawn, they will open. We use these passages to balance air pressure on the tympanic membrane. When we have an infection in the pharynx (throat), invaders can move through the passage and reach the middle ear region. The invaders will infect the ears at this time. This is why our ears often hurt, when we get an infection in the throat.

The pharynx is divided into three regions: nasopharynx, oropharynx and laryngopharynx. The nasopharynx is the superior part of the throat, close to the nasal cavity. The nasopharynx is posterior to the soft palate. The soft palate and uvula are what you see in the back of your throat. The soft palate will elevate when you swallow, to keep materials from going up towards the nose. The soft palate will depress when you are moving air through the nasal cavity. The oropharynx is located directly posterior to the oral cavity and the laryngopharynx is down close to the larynx (voice box). The pharynx is a common passageway for air, food and drink. This will place the pharynx in the respiratory system and digestive system.

LARYNX

Where the pharynx ends the larynx and esophagus begin. We want air to travel down the larynx (voice box). The esophagus will be looked at in the digestive system. The larynx is composed of several cartilages. The largest is the thyroid cartilage and it is this large cartilage, we see on the anterior surface of the larynx. Inside of the thyroid cartilage are the vocal cords. The vocal cords are two elastic tendons, stretched over the opening of the trachea (wind pipe). As air rushes past these tendons, they vibrate. The vibration of the tendons is primarily what produces the sound of the voice.

Think of when a guitar string is vibrated, it produces sound. Our vocal cords work in a similar way. On either side (laterally) of the vocal cords are the vocal folds. The folds connect to the side of the larynx and will come together to cover the larynx when swallowing. The epiglottis will assist by moving in an inferior direction to help cover the larynx when swallowing. So the vocal folds and the epiglottis prevent unwanted materials from entering the larynx. The opening between the vocal cords is the glottis.

The arytenoid cartilages attach to the vocal cords and through the use of skeletal muscles, the vocal cord length can be changed. When the cords are long, the sound of the voice is deeper and when the cords are shorter, the sound of the voice is higher. Testosterone causes growth of the larynx and this lengthens the vocal cords. This is why males have a deeper sound to their voices.

At the opening of the larynx is a piece of elastic cartilage called the epiglottis. The epiglottis is a cover for the larynx, when we swallow food or drink. When we are breathing the epiglottis is in a superior position, so that the larynx is open and air can pass.

Except for the epiglottis the remaining cartilage of the larynx is hyaline cartilage. This strong tissue is not there for protection, it is there to keep the air passageways open. We must keep these passageways open, because we are always moving air through them.

TRACHEA

The trachea is what we think of as our wind pipe. The trachea extends from the larynx to the primary bronchi. The trachea has strong, hyaline cartilage rings all the way down it. This tracheal cartilage is not there for protection, it is there to maintain an open airway. Just deep to this cartilage is smooth muscle. This trachialis smooth muscles is there for constriction. We will constrict this muscle when we cough or sneeze. The narrowing of the air passage ways will cause a more rapid movement of air.

Lining the inside of the trachea is a ciliated, pseudostratified layer of epithelial tissue. The cilia are moving mucous and all of the materials in it in a superior direction. As we move air through the bronchi (air passageways), we bring in foreign materials. This material will stick to the mucous and we must remove it. The cilia will keep this material moving up and out of the bronchi. If this material would remain, it would cause infection and inflammation. This would narrow the bronchi and inhibit air movement. The cilia and mucous form what is called the "muco-ciliary escalator".

If someone smokes for years, they can lose this cilia layer. If someone is always putting hot smoke and harmful chemicals in the bronchi the body will adapt. The ciliated pseudostratified layer will be replaced by a stratified squamous layer to keep the chemicals out of the body. This stratified layer will help do that, but the cilia are lost. Without the cilia the mucous pools and constricts the air ways. Smokers remove this mucous by coughing and the constant coughing damages the alveoli.

At the inferior point of the trachea, you run into a piece of cartilage called the carina. The carina has touch receptors all over its surface. If anything other than air touches these receptors, they will stimulate a strong cough reflex. When we accidentally take liquids or some other materials down the wind pipe, we cough uncontrollably because of these sensory receptors.

TRACHEOBRONCHIAL TREE

CONDUCTING ZONE

The tracheobronchial tree includes the trachea and all of the 23 branches of bronchi (air passageways). This series of passageways is called the tracheobronchial tree. This tree is divided into two zones, conducting and respiratory.

The conducting zone begins with the trachea and continues down to the terminal bronchioles. This zone ends here, because this is where the tracheal cartilage ends.

The trachea divides into the two primary bronchi. Each primary bronchus leads out to a lung. The primary bronchus divides to form the secondary bronchus. Each secondary bronchus leads to a lobe of the lungs. The secondary bronchus divides into tertiary bronchus. Each tertiary bronchus leads to a bronchopulmonary segment within the lungs. The bronchi enter and leave the lungs at a region called the hilum.

The conducting zone has the following characteristics.

1. It begins at the trachea and ends at the terminal bronchioles.
2. The first 16 of the 23 branches in the tracheobronchial tree are in this zone.
3. Cilia are found lining the inside of this zone.
4. Cartilage is found throughout this zone, holding the air passageways open.
5. There is no gas exchange in this zone; it is a passageway for air.

RESPIRATORY ZONE

The respiratory zone is the lower zone of the tracheobronchial tree. This zone begins where the conducting zone ends. The air passageways change as we go deeper into the lungs.

The respiratory zone has the following characteristics.

1. It begins at the terminal bronchioles and ends at the alveoli.
2. The last 7 of the 23 branches in the tracheobronchial tree are in this zone.
3. Macrophages are removing foreign materials in this zone instead of cilia.
4. No cartilage is found in this zone.
5. Gas exchange occurs in this area, because alveoli are found throughout this zone.

Without the cartilage in this zone, there is nothing to hold the air passageways open. If a person has an asthma attack, there is nothing to prevent the smooth muscle from constricting and narrowing the air passageways. It is this zone that will give asthma victims problems. If these are passageways constrict, air movement can be greatly reduced and gas exchange may be inadequate. Many people die from asthma attacks every year.

ALVEOLI

Alveoli are found along the respiratory zone and they are also found at the end of the zone in clusters. Alveoli are where gas exchange occurs in the lungs. The more alveoli we have, the more gas exchange we have. The thin layer between the blood in the pulmonary capillaries and the air in the alveoli is called the respiratory membrane. The respiratory membrane is made of two simple squamous epithelial cell layers. One layer is found in the wall of the alveoli and the other is the endothelium of the capillary. So two simple squamous epithelial layers are all that separate the blood from the air in the lungs.

The wall of an alveoli is made of two cell types. Both are epithelial cells, but with different functions. The wall is made of type I pneumocytes and type II pneumocytes. The type I pneumocytes are simple squamous epithelial cells and they cover 90% of the alveolar wall. These cells are there for gas exchange. The type II pneumocytes are simple cuboidal epithelial cells and they cover 10% of the alveolar wall. These cells are there to produce surfactant. Surfactant is needed on the inner alveolar wall to decrease the attraction of water molecules. All water molecules are like magnets, because of the polar covalent chemical bonding, which forms each one. Every time we exhale, these water molecules on the inner alveolar walls get closer to each other. The closer they get, the more they are attracted to each other. If we didn't have the surfactant to decrease this attraction, the water molecules would come together and collapse the alveolar walls, every time we exhaled. It would be like we had the wind knocked out of us, every time we exhaled. Everyone knows when you have the wind knocked out of you; it is difficult to reinflate the lungs. This is difficult because the water molecules are attracted to each other and don't want to let go. We must have the surfactant to prevent this strong attraction.

Type I pneumocytes
Cover 90% of alveolar walls
Are used for gas exchange
Are simple squamous epithelial cells

Type II pneumocytes
Cover 10% of alveolar walls
Are used for surfactant production
Are simple cuboidal epithelial cells

The alveoli are surrounded by elastic fibers. When we bring air into our lungs, the alveoli inflate like balloons. When the alveoli inflate, the elastic fibers are stretched. Stretching these fibers takes a lot of energy. When we relax the muscles of inspiration, these elastic fibers recoil and this recoil is largely what moves air out of our lungs. Think of it this way. When we bring air into our lungs these elastic fibers stretch like rubber bands. Everyone knows rubber bands recoil. When we relax the diaphragm, the elastic fibers recoil and move air out of our lungs. Think about a person getting CPR. Someone has to blow the air into the person's lungs, but the person giving CPR doesn't have to suck the air out. Why? The elastic fibers are stretched when the lungs are inflated. When someone stops blowing into the lungs, the elastic fibers recoil and force the air out.

Along with the elastic fibers, the water molecules help to move air out also. Remember that water molecules are like magnets. When we relax the lungs, these magnets are attracted to each other. As they are attracted to each other, they also help to deflate the lungs. Surfactant keeps this attraction from being too strong.

LUNGS

The lungs rest on top of the diaphragm muscle, which is the primary muscle of ventilation. The broad base of the lungs is on the inferior surface and the pointed apex is at the superior surface. The hilum is found medially and is the area where the bronchi, arteries and veins enter and exit the lungs.

The two lungs are not identical and this is because of the position of the heart. 2/3 of the heart is left of the sternum. Because of its position the left lung is smaller than the right lung. You will see the following differences in the lungs.

1. The left lung is smaller than the right.
2. The left lung has 2 lobes, while the right one has 3.
3. The left lung has 9 bronchopulmonary segments, while the right lung has 10.

The lungs are found inside the thoracic cavity and each individual lung is found inside of a pleural cavity. Each pleural cavity has a visceral and parietal membrane. The visceral layer is always the inner layer and will be the surface of the lung. The parietal layer is always the outer layer and is the inner surface of the thoracic cavity. Between these two layers we find pleural fluid. The two membranes and the fluid always have the same two functions, reducing friction and holding the organs in place.

MUSCLES OF VENTILATION

The muscles of ventilation are divided into two groups, the muscles of inspiration and the muscles of expiration.

The muscles of inspiration are the diaphragm, external intercostals, scalenes and pectoralis minor. The diaphragm is the prime mover in ventilation. When this muscle is relaxed, it is superior and dome shaped. When it contracts it moves down in an inferior direction. When the other inspiratory muscles contract they will take the ribs and swing them up and out. Think about how your thoracic cavity elevates, when you take a deep breath. All of these muscles are working to increase the size of the thoracic cavity. When we increase the size of this cavity, the pressure on the inside of it drops. As the interior pressure drops, air moves from outside our body, to inside our lungs. Air only moves when we have a pressure gradient. A pressure gradient exists when we have two areas of two different pressures and air always moves from the area of high pressure to low pressure.

    Muscles of inspiration
    Diaphragm, external intercostals, scalenes and pectoralis minor

    Muscles of expiration
    rectus abdominis and internal intercostals

The muscles of expiration are the rectus abdominis and internal intercostals. The rectus abdominis muscles will compress the organs in the abdominal cavity. When these organs are compressed they push up on the diaphragm muscle and when the diaphragm moves in a superior direction are moves out of the lungs. You can tell the diaphragm has been working, if you haven't ran in a long time. In the spring you may exercise and the next day, the stomach may be sore. This soreness is caused by the use of the abdominal muscles during expiration. At the same time the internal intercostals will move the ribs in an inferior direction. These muscles will make the thoracic cavity smaller. When the volume of the thoracic cavity decreases, the pressure on the inside increases. This increase in pressure will cause the air to move out of our lungs, where pressure is lower.

    When we are resting, we only need the diaphragm muscle to move air in and out of the lungs. When the diaphragm contracts, it moves in an inferior direction and draws the air in. When the diaphragm relaxes the recoil of the elastic fibers around the lungs and the attraction of the water molecules in the alveoli will move the air out. So we only need the diaphragm when resting.

    During labored breathing, such as times of exercise, we are

using all of the muscles of ventilation. This will move air in and out of the lungs as fast as possible and give greater gas exchange.

PLEURAL CAVITY

The lungs are inside the thoracic cavity and the each lung is separated by the mediastinum. The mediastinum is not a body cavity, but a separation between the lungs. Each lung is enclosed in a pleural cavity and like other body cavities, they have two membranes. The visceral pleura is the inner membrane and is the same as the surface of the lung. The parietal pleura is the outer membrane and is the same as the inner surface of the thoracic cavity. In between these two membranes is pleural fluid and the membranes and fluid always serving two purposes: reducing friction and holding the organs in place. The lungs are constantly inflating and deflating. Without something to reduce the friction, damage would occur. You can see how friction is reduced by imagining two pieces of glass. If you have two pieces of glass, one on top of the other. If you rub the two pieces together, they will scratch. Take the same two pieces of glass and put a little water between them, and rub them again. They will slide over each other easily and not scratch. The fluid between the lungs and thoracic wall reduce friction the same way. If you take the same two pieces of glass without the water, you can pull them apart easily, but with water in between them you can't. The fluid will also hold the organs in place.

BOYLES LAW

Boyle's Law tells us that there is an inverse relationship between pressure and volume. What that means is this, when volume is increasing, pressure is decreasing. The opposite applies also, when volume is decreasing, pressure is increasing. We use Boyle's Law to see how the muscles of ventilation move air in and out of our lungs.

When looking at ventilation, you must always consider two pressures: alveolar pressure - the pressure inside the lungs and barometric pressure – the pressure around our body. We can't change the barometric pressure around our body, but we can change the alveolar pressure within our lungs. By changing the pressure in our lungs, we move air in and out of them. Remember, for air to move there must be a pressure gradient, meaning two areas of different pressure and air always moves from high to low pressure.

Let's start with inspiration. When the diaphragm muscle contracts, it moves down in an inferior direction. Since the diaphragm is the inferior border of the thoracic cavity; when this muscle moves down, the thoracic cavity gets bigger. At the same time other muscles can swing our ribs up and out. This movement will also make the thoracic cavity bigger. Getting bigger is the same as increasing the volume. If the volume is increasing, then the pressure is decreasing. Remember volume and pressure are always opposite of each other. So when the diaphragm contracts the pressure inside the thoracic cavity drops. If the pressure inside our chest drops the air from around our body will move in to the lungs. Remember air always moves from high to low pressure, so if we suddenly drop our thoracic pressure (inside pressure), the air comes in. This is how we fill our lungs with air. Notice that nothing is forcing the lungs open. We just increase the volume (size) and this gives us a pressure change. This is also assisted by the thoracic wall and the fluid in between it and the lungs. Remember that the thoracic wall and the lungs have fluid in between them. This water helps to hold the structures together, so when the thoracic cavity gets larger, it will help to move the lungs with it and inflate them.

Let's look at expiration. When the diaphragm muscle relaxes, it moves up in a superior direction. This movement will make the thoracic cavity smaller. Other muscles can pull down on the ribs at the same time, also making the cavity smaller. The cavity getting smaller is the same as saying the volume is decreasing. If the volume is decreasing, then the pressure is increasing. If we suddenly increase the pressure inside our lungs, then the air moves out towards the lower pressure. Don't forget we can't change the barometric pressure around our body, but we can change the internal pressure. By changing the pressure, we establish a pressure gradient and air will move from high to low pressure.

The expiration of air is largely accomplished by the recoil of elastic fibers around the alveoli. When the muscles of inspiration contract the alveoli expand and this stretches the elastic fibers. When the muscles relax the elastic fibers recoil and help to force air out of the lungs.

The water on the inside of the alveoli also helps to move air out of the lungs. All water molecules are like little magnets, so they are attracted to each other. The muscles of inspiration pull these little magnets apart, during inspiration. Then when those muscles relax, the attraction of the water molecules pulls the alveolar walls closer together. As the alveolar volume decreases, air moves out of the lungs. Surfactant reduces the attraction of the water molecules. Without this surfactant, it would be very difficult to inflate our lungs.

Compliance is how easy our lungs stretch. Compliance is influenced by the elastic fibers and the water inside the alveoli. The elastic fibers and water decrease the compliance of the lungs, but not so much as to cause a problem. But if we have too much water inside the alveoli or the elastic fibers are replaced by another fiber, we may decrease our compliance. When compliance decreases, it becomes more difficult to inflate the lungs and this can result in inadequate gas exchange. Emphysema decreases compliance and causes ventilation problems.

Around the seventh month of development, our lungs begin producing surfactant. If a baby is born before this seventh month, they can have very big lung problems. Without the surfactant production, the water attraction will be to strong. This will cause great difficulty during inspiration and the individual could die. This is called hyaline membrane disease of the newborn.

When looking at air movement, the same rules apply that we used with blood movement. We must consider the resistance to air movement, just like we did with blood. Remember that the most important variable with resistance is radius. A small change in radius, gives us a big change in resistance. We can change the radius of our bronchi, by constricting the smooth muscle found inside of the air passageways. When this smooth muscle constricts, the radius of the air passageways get smaller. This decrease in radius will greatly increase resistance to the air flow. When resistance increases, air flow decreases. This is what happens when someone has an asthma attack. The respiratory zone doesn't have cartilage to keep the bronchi open. When the smooth muscle inside the bronchi constricts the air passageways gets to small, air flow decreases and the individual can die.

PULMONARY VOLUMES

A spirometer is a device used to measure the volume of air moving in and out of the lungs. A few measurements are common and need to be understood.

Tidal volume – the volume of air moved in and out of the lungs with a normal resting breath. This volume comes to about 500ml of air.

Expiratory reserve – the volume of air which can be exhaled after a normal breath. This volume comes to about 1100ml.

Inspiratory reserve – the volume of air which can be inhaled after a normal breath. This volume comes to about 3000ml.

Residual volume – the volume of air remaining in the lungs, no matter how hard someone exhales. This volume comes to about 1200ml.

Total lung capacity – the sum of the four previous volumes.

Vital capacity – the maximum amount of air someone can move in and out of the lungs. This is when a person takes a very deep breath and then blows out as much air as possible. This amount will vary greatly. Factors such as height, weight, age, physical condition and others affect vital capacity.

Minute ventilation – how much air we move in and out of the lungs in one minute. This is easy to calculate; all we need to do is multiply respiratory rate X tidal volume. Respiratory rate is how many times a person breathes in one minute. The average person breathes 12-14 times a minute. The tidal volume we looked at above. Multiply 12 X 500ml and this is minute ventilation. This comes to 6 liters of air per minute.

When we look at the volume of air moved in and out of our lungs, this volume doesn't represent the volume of air used for gas exchange. Much of this air is not in the alveoli and air must be in the alveoli for gas exchange to occur. Much of the air is found in the nose, pharynx, larynx, trachea and bronchi (conducting zone). This air is not available for gas exchange. The area containing this air is considered the anatomical dead space. If any alveoli are lost, this will add to the dead space. This additional dead space is called physiologic dead space. Hopefully you don't add any physiologic dead space to your lungs.

DALTON'S LAW

Dalton's Law states that the total pressure of a gas mixture is the sum of the individual pressures of each individual gas. It is easy to calculate the individual pressure of a gas in a mixture. We need a starting point for gas pressure and we will use barometric pressure at sea level. At sea level the pressure around your body is 760mmHg. What we are breathing is primarily oxygen (21%) and nitrogen (79%). To calculate the pressure each gas exerts in a mixture, we just multiply the percent times the pressure. So the partial pressure of oxygen is .21 X 760 = 160mmHg. The pressure exerted by each individual gas is the partial pressure of the gas. This is significant because gases move from areas of high partial pressure, to areas of low partial pressure. These partial pressures will determine where gases will diffuse and how rapidly they will diffuse. The greater the partial pressures the faster diffusion will occur.

Diffusion rates are affected by many variables. Factors affecting diffusion are:

   1. The greater the difference in partial pressures the faster

the rate of diffusion.

2. If the respiratory membrane thickens, diffusion will decrease. If fluid builds up inside the lungs, then gases must diffuse further, this will decrease diffusion.

3. If surface area decreases, diffusion will decrease. If alveoli are lost for any reason, we will have less surface area for gas exchange. This loss will cause fewer gases to be swapped.

OXYGEN TRANSPORT

Almost all oxygen is transported by the hemoglobin molecules of the red blood cells. Variables within the body influence how much oxygen this hemoglobin is transporting. pH is one big variable which influences oxygen transport. Our pH must stay in the range of 7.35-7.45. If we leave this pH range, the proteins in our body change shape. When they change shape they can no longer do their job. If hemoglobin moves into an acidic environment, the proteins change shape and will no longer bind to and transport oxygen. So pH is a big factor in oxygen transport.

Our body will use this pH change to manipulate proteins in the body. Think about this, "How do our hemoglobin molecules know when to pick up or release oxygen?" They don't, but the body can use pH and temperature to control them. When hemoglobin moves into a tissue like skeletal muscle, the hemoglobin moves into a low pH environment. Living cells are releasing carbon dioxide and as we get more carbon dioxide, we get more hydrogen ion. This will create an acidic environment. As the hemoglobin moves into an acidic environment, the proteins change shape and this is when they release the oxygen. As we move this same blood back to the lungs, carbon dioxide is released and this eliminates the excess hydrogen ions. The pH will return to normal and the hemoglobin goes back to its normal shape, this is when the hemoglobin picks up oxygen. So pH changes greatly influence the function of hemoglobin.

Our body also uses temperature to manipulate proteins in the body. Think about what happens when hemoglobin moves into exercising skeletal muscle. Everyone know when we exercise, we heat up. When hemoglobin moves into this warm environment, the proteins change shape. When the hemoglobin changes shape they release the oxygen. This is what we want hemoglobin to do, when it moves into exercising skeletal muscle. So the body uses temperature to manipulate hemoglobin molecules.

CARBON DIOXIDE TRANSPORT

Hemoglobin moves practically all oxygen, but it only moves about ¼ of all carbon dioxide. Most carbon dioxide is moved around the body in the form of bicarbonate ion. This is because of the chemical reaction we have mentioned before.

$$CO_2 + H_2O \leftrightarrow H_2CO_3 \leftrightarrow H + HCO_3$$

When carbon dioxide is released, it quickly becomes bicarbonate ion. About 70% of all carbon dioxide is moved around the body in the form of bicarbonate ion.

RESPIRATORY CONTROL

The pons and medulla oblongata within the brainstem have control over ventilation. It is not clearly understood how they work together, but if damage occurs to these areas, ventilation can be hindered.

The cerebrum is where we have conscious thought and obviously we can control the muscles of ventilation with our conscious thought. We can hold our breath when we want, like when we dive underwater.

We have chemoreceptors for oxygen and carbon dioxide located within the medulla oblongata, as well as the aorta and internal carotids. As oxygen levels decrease, we see an increase in ventilation. Hypoxia is the condition of having too little oxygen in the blood. As carbon dioxide levels increase, we see an increase in ventilation. If these chemicals move in the opposite direction, we would see ventilation decrease. Hypercapnia is the condition of having too much carbon dioxide in the blood and hypocapnia is the condition of having too little carbon dioxide in the blood.

Pain and the sympathetic division of the nervous system can regulate ventilation. The sympathetic division of the nervous system will increase ventilation, when we become physically active or when we feel pain.

The Hering-Breuer reflex involves baroreceptors in the lungs. These stretch receptors will prevent us from over inflating the lungs.

RESPIRATORY DISORDERS

1. COPD – chronic obstructive pulmonary disorder – This is not one disorder, but will include anything causing a decrease in pulmonary function over time.

2. Hyaline membrane disease – respiratory distress in a newborn, caused by a lack of surfactant.

3. Pneumothorax – air in the thoracic cavity, due to a hole in the thoracic cavity. Air will fill the space between the lung and thoracic wall, instead of going into the lung.

4. Hyperventilation – a rapid respiratory rate. This will result in high oxygen levels and low carbon dioxide levels.

5. Hypoventilation – a depressed respiratory rate. This will result in low oxygen levels and high carbon dioxide levels.

6. Apnea – the temporary loss of breathing.

7. Hypercapnia – high carbon dioxide levels.

8. Hypocapnia – low carbon dioxide levels.

9. Hypoxia – low oxygen levels.

Chapter 6 – Study Questions

1. The movement of air in and out of the lungs is?
a. respiration
b. ventilation
c. transpiration
d. oxidation
e. reduction

2. The swapping of gases is?
a. respiration
b. ventilation
c. transpiration
d. oxidation
e. reduction

3. The swapping of gases in the lungs is?
a. ventilation
b. internal respiration
c. external respiration
d. transpiration
e. external ventilation

4. The swapping of gases at the tissues is?
a. ventilation
b. internal respiration
c. external respiration
d. transpiration
e. external ventilation

5. Carbon dioxide will combine with water to form?
a. bicarbonate ion
b. hydrogen ion
c. carbonic acid
d. hydrochloric acid
e. hemoglobin

6. When carbon dioxide levels increase, hydrogen ion levels will?

a. increase
   b. decrease
   c. no effect

7. When carbon dioxide levels decrease, hydrogen ion levels will?
   a. increase
   b. decrease
   c. no effect

8. When carbon dioxide levels increase to abnormal levels, a person will develop what condition?
   a. acidosis
   b. alkalosis
   c. both conditions
   d. neither conditions

9. When carbon dioxide levels decrease to abnormal levels, a person will develop what condition?
   a. acidosis
   b. alkalosis
   c. both conditions
   d. neither conditions

10. Too many hydrogen ions in the blood will give what condition?
    a. acidosis
    b. alkalosis
    c. both conditions
    d. neither conditions

11. Too few hydrogen ions in the blood will give what condition?
    a. acidosis
    b. alkalosis
    c. both conditions
    d. neither conditions

12. Normal pH range for our blood is?
a. 7.00-8.00
b. 7.50-7.60
c. 7.20-7.30
d. 7.35-7.45
e. 7.10-8.10

13. The respiratory system begins where?
a. pharynx
b. oral cavity
c. nasal openings
d. larynx
e. alveoli

14. The respiratory system ends where?
a. pharynx
b. oral cavity
c. nasal openings
d. larynx
e. alveoli

15. The nasal cavity functions to?
a. warm air
b. clean air
c. humidify air
d. sense of smell
e. all of the above

16. The auditory tubes function to?
a. transport gases to the lungs
b. move air to the paranasal cavities
c. clean air
d. balance air pressure
e. warm air

17. Our wind pipe is the same as?
a. pharynx
b. larynx

c. trachea
d. bronchi
e. alveoli

18. Our voice box is the same as?
a. pharynx
b. larynx
c. trachea
d. bronchi
e. alveoli

19. Our throat is the same as?
a. pharynx
b. larynx
c. trachea
d. bronchi
e. alveoli

20. The conducting zone of the tracheobronchial tree begins where?
a. pharynx
b. larynx
c. trachea
d. bronchi
e. alveoli

21. The conducting zone uses what structures to remove foreign materials?
a. cilia
b. microvilli
c. macrophages
d. flagella
e. none of the above

22. The air passageways are held open by what?
a. elastic cartilage
b. hyaline cartilage
c. fibrocartilage

d. bone
e. none of the above

23. Gas exchange only occurs at the?
a. pharynx
b. larynx
c. trachea
d. bronchi
e. alveoli

24. The function of a type I pneumocyte is?
a. gas exchange
b. surfactant production
c. to hold the bronchi open
d. red blood cell production
e. protection

25. The function of a type II pneumocyte is?
a. gas exchange
b. surfactant production
c. to hold the bronchi open
d. red blood cell production
e. protection

26. Surfactant is used to?
a. hold the bronchi open
b. reduce the attraction of water molecules
c. produce white blood cells
d. control the muscles of inspiration
e. none of the above

27. Which lung is larger?
a. right
b. left
c. neither

28. The cartilage which covers the larynx when swallowing is the?

a. thyroid cartilage
b. epiglottis
c. glottis
d. all 3
e. none of the above

29. Which is a muscle of inspiration?
a. rectus abdominis
b. internal intercostals
c. diaphragm
d. rectus femoris
e. temporalis

30. The two pleural membranes and the pleural fluid will provide what functions?
a. reduce friction
b. holds organs in place
c. a and b
d. neither

31. Boyle's Law tells us that?
a. when volume increases pressure increases
b. when volume increases pressure decreases
c. when volume decreases pressure decreases

32. When the diaphragm muscle contracts?
a. the thoracic cavity increases in volume and the pressure inside decreases.
b. the thoracic cavity decreases in volume and the pressure inside decreases.
c. the thoracic cavity increases in volume and the pressure inside increases.

33. When does surfactant production begin?
a. 3rd month of development
b. 5th month of development
c. 7th month of development
d. 9th month of development

e. after birth

34. Surfactant is needed in the lungs for?
a. lung development
b. decrease the attraction of water molecules
c. red blood cell production
d. emphysema treatment
e. trachea development

35. If the radius of the bronchi decreases, air flow will?
a. increase
b. decrease
c. no effect

36. The volume of air moved in and out of the lungs, during normal resting breathing is?
a. tidal volume
b. inspiratory reserve
c. expiratory reserve
d. residual volume
e. total lung capacity

37. No matter how hard we exhale, there is always air left in the lungs. This is called?
a. tidal volume
b. inspiratory reserve
c. expiratory reserve
d. residual volume
e. total lung capacity

38. The volume of air moved in and out of the lungs, during normal resting breathing is?
a. 200 ml
b. 400 ml
c. 500 ml
d. 700 ml
e. 1000 ml

39. As the difference in partial pressures increases, diffusion rates will?
a. increase
b. decrease
c. no effect

40. Most carbon dioxide is moved around the body in what form?
a. carbon monoxide
b. bicarbonate ion
c. carbonic acid
d. water
e. none of the above

41. Ventilation can be regulated by the?
a. pons
b. medulla oblongata
c. cerebrum
d. all of the above
e. none of the above

42. Which is the primary regulator of ventilation?
a. oxygen
b. carbon dioxide
c. carbonic acid
d. bicarbonate ion
e. none of the above

43. The smallest of these structures is?
a. pharynx
b. larynx
c. trachea
d. bronchi
e. alveoli

Answers to multiple choice questions.

1. B
2. A
3. C
4. B
5. C
6. A
7. B
8. A
9. B
10. A
11. B
12. D
13. C
14. E
15. E
16. D
17. C
18. B
19. A
20. C
21. A
22. B
23. E
24. A
25. B
26. B
27. A
28. B
29. C
30. C
31. B
32. A
33. C
34. B
35. B
36. A
37. D

38. C
39. A
40. B
41. D
42. B
43. E

DIGESTIVE SYSTEM

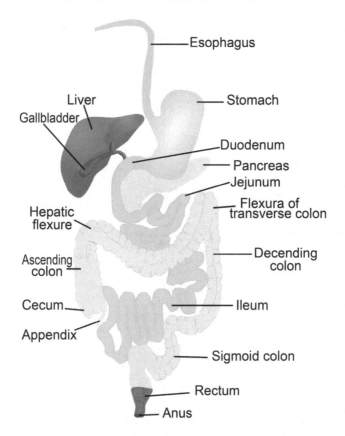

CHAPTER 7
DIGESTIVE SYSTEM

The digestive system is largely a tube extending through the body from the mouth to anus. Sometimes the digestive system is also called the GI tract or alimentary canal, but these terms are not the same. The GI tract and alimentary canal don't include all of the organs of the digestive system. The alimentary canal is the tube going through the body, but it doesn't include the accessory organs of the digestive system. The accessory organs are the organs of this system that food doesn't pass through. The salivary glands, liver, gallbladder and pancreas are all accessory organs of this system.

The organs of the digestive system are the oral cavity, salivary glands, pharynx, esophagus, stomach, small intestine, liver, gallbladder, pancreas, large intestine, appendix, and anus. This system includes many organs and takes up much of the abdominopelvic cavity.

GI TRACT LAYERS

The GI tract has four layers and these layers can be seen in most of the system.

1. Mucosa layer – This is the inner layer, which comes in contact with the food. This layer is covered with mucous and this mucous helps to liquefy the food and lubricate the GI tract. The epithelial tissue covering the inside is stratified squamous is some areas and columnar in others. Along the small intestine a large number of lymph nodes can be found. We absorb most materials in the small intestine, so a large number of lymph nodes are needed for filtration of foreign materials.

2. Submucosa layer – The submucosa is the layer superficial to the mucosa. This layer is highly vascular, so as to absorb materials. Many glands are found in this layer, adding materials to the mucosa when needed.

3. Muscularis layer – This is a thick layer of muscle along with the enteric plexus. The muscle will move materials along the GI tract and the enteric plexus will control the muscle and glands.

4. Serosa – The serosa is the most superficial layer and is the same as the visceral peritoneum. When looking at the surface of the organs, this is the layer seen. In places this visceral peritoneum meets the parietal peritoneum. Where they meet we find mesenteries. If you have ever seen someone hold up the internal organs of an animal or perhaps a cadaver, you may have seen the mesenteries. If you hold up intestines and see the thin layers of tissue holding the structures together, these are the mesenteries. The mesenteries hold the organs in place and make a pathway for blood vessels and nerves. Some of the more noticeable mesenteries are the greater and lesser omentum. The greater omentum forms a pouch called the fatty apron, because fats accumulate in it.

Some organs of the digestive system are outside of the parietal peritoneum. These organs are called retroperitoneal organs. The pancreas, duodenum, kidneys, parts of the colon and rectum are retroperitoneal.

ORAL CAVITY

The oral cavity is the beginning of the digestive system. The oral cavity is largely about mechanical digestion. Mechanical digestion involves the movement of body parts breaking down food materials. The mouth contains 32 teeth. We have 16 teeth in the mandible and 16 in the maxilla. The mandible and maxilla are separated into two halves, giving us four quadrants. Each quadrant of the oral cavity contains 8 teeth: 2 incisors, 1 canines, 2 premolars and 3 molars. The incisors and canines are for biting and tearing. The premolars and molars are for grinding. The proximal ends of the teeth are surrounded by dense, collagenous, connective tissue called gingiva (gums). A small amount of chemical digestion occurs due to lipases and amylases in the oral cavity.

The oral cavity is bordered by the lips in the anterior region, the cheeks at the lateral regions, the hard palate superior and the soft palate posterior. The uvula hangs off the soft palate and can be seen in the rear of the oral cavity. The soft palate will elevate when we swallow to prevent food from going up toward the nasal cavity. At the rear of the oral cavity is the opening to the pharynx called the fauces. Lateral to the fauces the palatine tonsils can be seen.

The muscles of mastication (chewing) are the masseters, temporalis, lateral pterygoids and medial pterygoids.

The tongue is used to manipulate food over the teeth. The tongue is largely skeletal muscle. This is why we have conscious control over it. The tongue is secured anteriorly by the lingual frenulum, which can be seen when elevating the tongue. Lipid soluble drugs are sometimes delivered under the tongue, these are the sublingual drugs. This is done because the epithelial layer superficial to the capillaries is thin in this area. Placing a lipid soluble drug here will get it into the circulation rapidly.

The oral cavity contains three sets of salivary glands. The submandibular glands are beneath the mandible, the sublingual are beneath the tongue and the parotids are anterior to the ears. The parotid salivary glands become greatly inflamed when someone has the mumps.

Saliva plays an important role in oral hygiene. The constant production of saliva washes bacteria out of the oral cavity. Without this washing action, we would have many more infections of the teeth and gingiva. The saliva also helps to liquefy the food and soften it. The saliva contains a few enzymes, which begin the chemical digestion of food, but this is a very small amount. After food has been chewed, it is then called a bolus.

PHARYNX

After chewing food we will use the tongue to push food up against the hard palate and then back. The soft palate will move in a superior direction and the bolus moves back into the pharynx (throat). The pharynx has three sets of pharyngeal constrictor muscles: a superior, middle and inferior set. When swallowing (deglutition) the muscles will work from superior to inferior and this will move the food down in an inferior direction.

ESOPHAGUS

When anything reaches the inferior part of the pharynx, there are two possible routes. The trachea extends toward the lungs and the esophagus extends towards the stomach. When food approaches the trachea the epiglottis will cover the trachea and food will move towards the esophagus. The esophagus has a sphincter (round) muscle at its origin. This is the upper esophageal sphincter and this muscle will relax to allow materials into the esophagus. The esophageal muscles will contract one after another to move food down the esophagus. The rhythmic movement of this muscle is called peristalsis.

The esophagus passes through the diaphragm muscle to reach the stomach in the abdominal cavity. Where the esophagus passes, it leaves a weak spot in the diaphragm called the esophageal hiatus. Sometimes a structure, such as small intestine, finds its way up through this opening and gets constricted. Blood flow could be cut off and the tissue could die. This is called a hiatal hernia. At the inferior part of the esophagus is the lower esophageal sphincter. This muscle will relax to allow materials to enter the stomach.

STOMACH

The esophagus meets the stomach at the lower esophageal sphincter. This muscle will contract to keep materials in the stomach. Sometimes this muscle is unable to do so and the acidic contents of the stomach move back up into the esophagus. This is the cause of heart burn.

The superior part of the stomach is the fundus. The fundus is also the broadest part and the only part superior to the lower esophageal opening. Where the esophagus meets the stomach, this area is called the cardiac region. The heart is close to this area, thus the name. Because of the close position of the heart, people sometimes confuse bad heartburn with a heart attack and sometimes confuse a mild heart attack with bad heart burn. The body is the main part of the stomach and extends down to the pyloric region. The pyloric region contains another circular smooth muscle called the pyloric sphincter. This muscle regulates what can leave the stomach and enter the duodenum (first part of the small intestine). As the stomach curves it forms the lesser curvature on the superior surface and the greater curvature on the inferior surface. When the stomach is empty, rugae (wrinkles) can be seen on the inner surface. These folds allow for expansion, when the stomach fills with food. Baroreceptors will tell us when the rugae are unfolding, thus telling us when we are full.

The stomach contains gastric glands with several cell types within them. The cells of the stomach are:

1. Surface mucous cells – These cells produce mucous on the inner lining of the stomach. This mucous lining is very thick in the stomach and this thick layer will protect the stomach lining from acid and secretions.

2. Neck mucous cells – These cells are lining the ducts of the gastric glands with mucous.

3. Chief cells – Chief cells release several secretions, one of them is pepsinogen. Pepsinogen is used for the digestion of proteins.

4. Parietal cells – Parietal cells release hydrochloric acid and intrinsic factor. Intrinsic factor helps with the absorption of B12. Hydrochloric acid has a minor digestive effect; it's primarily for enzyme activation and killing bacteria. Acid release can drop the pH of the stomach to a 1, when food is present.

5. Endocrine cells – Like endocrine cells elsewhere, endocrine cells release hormones.

There is only a small amount of absorption in the stomach. The stomach is largely a mixing chamber for food. A large amount of water and secretions are added to the food and the materials are mixed. The stomach will make the food into a liquid material, so as to increase surface area. The smaller food materials are broken down, the more surface area the food has. Many students don't understand that when one large particle is broken down, the particles get smaller, but the collective surface area increases. The greater the surface area, the better chemical digestion will work, once the food materials reach the small intestine and its enzymes. This fluid food material in the stomach is called chyme.

SMALL INTESTINE

The small intestine is found just after the pyloric sphincter of the stomach. The stomach will empty its fluid contents into the duodenum of the small intestine and many other materials from the liver, gallbladder and pancreas will soon join it. The small intestine gets its name because it's small in diameter. The small intestine is much longer than the large intestine, so remember to think diameter when it comes to large and small intestines. The small intestine is about 21 feet long compared to the 6 feet of the large intestine.

The three sections of the small intestine are duodenum, jejunum and ileum. The duodenum is the first section and makes a 180 degree turn to bend back up under the stomach. In this bend the head of the pancreas can be found. The liver, gallbladder and pancreas will all release their materials into the duodenum. Between these organs and the stomach, a very large amount of materials enter the duodenum every day.

In the interior of the duodenum is the major and minor duodenal papilla. The liver and gallbladder will release their materials through the major papilla and the pancreas will release through both papilla. These papilla are sphincter muscles which will regulate the entrance of materials into the duodenum.

After the duodenum are the jejunum and then the ileum. The

duodenum is about 1 foot long, the jejunum 8 feet and the ileum is 12 feet. The small intestine gradually narrows in diameter and there are no clear barriers between the sections. Most materials are absorbed in the proximal end of the small intestine and less is absorbed toward the jejunum and ileum.

The primary function of the small intestine is nutrient absorption. Around 92% of all nutrients will be absorbed in the small intestine. The inner lining of the small intestine contains a large amount of microvilli. Microvilli are always used to increase surface area. Without this large number of microvilli, we couldn't absorb enough nutrients to survive. Because of all of the absorption, a large amount of foreign materials enter the body. Hundreds of lymph nodes will be filtering out the harmful materials to protect the body. Lacteals are the special lymphatic vessels found along the GI tract, which will absorb lipids. Lymph plus the lipids within is known as chyle.

LIVER

The liver is a large internal organ and consists of four lobes. When looking at the liver in the body, the large right lobe and the smaller left lobe can be seen. The gallbladder can be seen on the inferior surface of the right lobe. The caudate and quadrate lobes can be seen on the inferior surface and in between them is a porta. The porta is the area where the hepatic duct, hepatic artery and hepatic portal vein pass through the liver. The hepatic portal vein is a rare example of a vein leading into an organ. This portal vein returns materials from the GI tract, mostly from the small intestine, to the liver. There are several ducts visible on the liver. The hepatic duct will join the cystic duct of the gallbladder and form the common bile duct. The common bile duct carries bile from the liver and gallbladder to the duodenum. The common bile duct meets the pancreatic duct to form the hepatopancreatic ampulla.

Internally the liver is divided into lobules. The lobules are six sided structures made of many hepatic cords. At the center of each lobule is a large central vein with the cords radiating out in all directions. These cords look like the spokes on a bicycle wheel.

They radiate out from the center to the outer edges. Each hepatic cord is made of many hepatocytes (liver cells). Around the outer edges of the lobules are portal triads. Each triad is a hepatic duct, hepatic artery and hepatic portal vein.

The liver has 6 major functions.

1. Detoxification. As materials are absorbed by the GI tract, all of it enters the hepatic portal system and is taken to the liver and nowhere else. This portal system has been set up to ensure that absorbed materials aren't immediately taken to the general circulation. Some of what we absorb is harmful and it must be removed. The hepatic portal system ensures that absorbed materials will be taken to the liver first. The liver will remove foreign chemicals like drugs, alcohol, etc. The liver is a first line of defense against these materials. When studying pharmacology, you will discuss the first pass effect of oral medications. What this means is that if any oral medication is going to work, it must first pass the liver. The liver will remove around ½ of many oral medications, before they have a chance to go anywhere else in the body and work. Many foreign invaders will be removed by macrophages in the liver. The hepatocytes also remove ammonia from the blood and convert it to urea. Urea will then be removed by the kidneys.

2. Bile production. Bile is produced entirely by the liver and nowhere else. Bile is largely waste products from the blood. As the liver removes these wastes, it concentrates them into what we call bile. This bile is released into the small intestine and leaves the body. This is one form of waste removal.

Bile is largely a basic material, so it will help to neutralize the acids being released from the stomach. This acid material would prevent pancreatic enzymes from working properly. Bile is also needed for the breaking down of lipids. When lipids enter the duodenum bile will be released and the lipids will be broken down into smaller lipid pieces.

3. Storage. The liver stores up large amounts of energy in the form of glycogen. When we eat a meal the hepatocytes will increase in size as they store energy, later when we haven't eaten for

some time, the hepatocytes will decrease in size as they release energy. Liver cells change size through the day, depending on how much we eat and exercise. Vitamins and some minerals like iron are stored in the liver.

4. Synthesizing nutrients. The liver can take some materials from the blood; we have in abundance and convert them to some materials we are lacking. This is the interconversion of nutrients.

5. Synthesizing plasma proteins. Many of the proteins we find in the blood are synthesized by the liver.

6. Phagocytosis of old red blood cells, bacteria and other debris. As we lose a few million red blood cells every second, many of these cells are removed from the blood by the liver and the spleen.

GALLBLADDER

The gallbladder doesn't make anything. The gallbladder is only a muscular storage container for bile. As the liver makes bile, the bile moves back up the cystic duct and enters the gallbladder. This bile will be released when lipids pass through the small intestine. Remember bile is good for breaking down lipids into smaller particles. It isn't a chemical digestion, but it does break them down. If we eat more lipids than we have bile to break it down, materials may move through the GI tract to rapidly and diarrhea may be the end result. If someone has the gallbladder removed, they won't be able to process as many lipids as they could previously.

Gallstones are the result of cholesterol, calcium and bilirubin precipitating in the gallbladder. These stone like structures can block ducts and cause pain.

PANCREAS

The pancreas anatomy is simple. It has a head, which lies in the bend of the duodenum, a body and a tapered tail piece. The pancreas is a mixed gland, meaning it has endocrine and exocrine parts. The hormones of the endocrine cells were previously discussed. The exocrine part involves the digestive system. The pancreas has a large pancreatic duct through the middle of it. All exocrine glands have ducts leading to something and this one leads to the duodenum.

The pancreas produces the bulk of our digestive enzymes and it produces digestive enzymes for all of the major organic materials we consume. Lipids will be broken down by lipases, carbohydrates will be broken down by amylases, DNA will be broken down by deoxyribonuclease and proteins are broken down by several enzymes. The enzymes which break down organic materials are not difficult to remember. Any lipase will break down lipids, any amylase will break down carbohydrates, enzymes which break down DNA and RNA have DNA and RNA in their name. The remaining enzymes with odd names usually break down proteins. The pancreas secretions also contain a large amount of bicarbonate ions for neutralizing acids coming from the stomach.

LARGE INTESTINE

The primary function of the large intestine is the absorption of most of the remaining water. The chyme from the stomach will be converted into feces in the large intestine. If most of the remaining water is not reabsorbed diarrhea will result, if too much water is reabsorbed constipation will result.

The large intestine is named by its large diameter. It is shorter than the small intestine, so don't confuse diameter and length. The major regions of the large intestine are the cecum, colon, rectum and anal canal. Where the small intestine ends there is a sphincter muscle called the ileocecal valve. This muscle regulates what will leave the small intestine and enter the cecum. Extending off the cecum is the appendix, a small finger like structure. The function of the appendix is debated. Some believe it is vestigial, meaning it's left over from a previous organ and is no longer functional. Some believe the appendix acts as a reservoir for useful bacteria.

After the cecum the colon is the next part of the large intestine. The colon is divided into several sections. The first is the ascending colon, named because it travels from the lower right quadrant up to the liver. When reaching the liver the colon turns and this bend is called the right hepatic flexure.

The second part of the colon is the transverse colon. Trans means across and that is what this part of the colon does, it travels across the abdominal cavity. The transverse colon travels from the liver and moves to the upper left quadrant and meets the spleen. The colon will turn at the spleen and this bend is called the left splenic flexure.

The third part of the colon is the descending colon, named because it travels from the spleen in an inferior direction. The descending colon will travel down to the lower left quadrant. From here it will bend twice in as S shape making the sigmoid colon (fourth part). After this we find the rectum, which is a holding chamber for feces. Baroreceptors in the rectum will tell us when this chamber is full and this is when defecation is needed. This stretching of the rectum stimulates the defecation reflex.

Along the large intestine we see segments called haustra. These pouches are formed by the contraction of smooth muscle along the large intestine. Epiploic appendages are fat accumulations along the outer surface. Three long, thin muscle layers called teniae coli are found along the colon also. The movement of feces by the smooth muscle along the colon are called mass movements.

There are a large number of helpful bacteria in the large intestine. They provide us with a needed material like vitamin K and we provide them with a home. These bacteria produce a gas called flatus.

Chapter 7 – Study Questions

1. Which is not an accessory organ of the digestive system?
a. salivary glands
b. liver
c. stomach
d. gallbladder
e. pancreas

2. The deepest layer of the GI tract is the?
a. mucosa
b. submucosa
c. muscularis
d. serosa
e. none of the above

3. Which layer of the GI tract comes in contact with food?
a. mucosa
b. submucosa
c. muscularis
d. serosa
e. none of the above

4. The visceral peritoneum is the same as?
a. mucosa
b. submucosa
c. muscularis
d. serosa
e. none of the above

5. Mesenteries will?
a. hold organs in place
b. make a pathway for blood and nerves
c. both
d. neither

6.. The digestive system begins at the?
a. pharynx
b. esophagus

c. oral cavity
d. stomach
e. small intestine

7. The average adult has how many teeth?
a. 22
b. 30
c. 32
d. 40
e. 42

8. Which is not a muscle of mastication?
a. masseter
b. temporalis
c. lateral pterygoid
d. medial pterygoid
e. orbicularis oris

9. The flap of tissue dangling in the back of the oral cavity is the?
a. hard palate
b. tongue
c. temporalis muscle
d. uvula
e. tonsils

10. The salivary glands located anterior to the ears?
a. submandibular
b. parotid
c. sublingual
d. none of the above

11. Chewed food is called a?
a. bolus
b. chyme
c. feces
d. flatus
e. deglutition

12. Swallowing is also known as?
a. bolus
b. chyme
c. feces
d. flatus
e. deglutition

13. Food should not pass through what part of the pharynx?
a. nasopharynx
b. oropharynx
c. laryngopharynx

14. Which muscle will relax to allow food to enter the esophagus?
   a. pyloric sphincter
   b. lower esophageal sphincter
   c. upper esophageal sphincter
   d. major duodenal papilla
   e. gallbladder

15. Which muscle will contract to keep food contents in the stomach?
   a. pyloric sphincter
   b. lower esophageal sphincter
   c. upper esophageal sphincter
   d. major duodenal papilla
   e. gallbladder

16. The region where the esophagus meets the stomach is called the what region?
   a. cecal region
   b. pharynx region
   c. hepatic region
   d. cardiac region
   e. surface region

17. An empty stomach will have what on the inside?
a. flatus
b. rugae
c. fauces
d. frenulum
e. gingival

18. The broadest part of the stomach is the?
a. fundus
b. body
c. rugae
d. pyloris
e. duodenum

19. Hydrochloric acid is produced by which cells?
a. surface mucous
b. neck mucous
c. parietal
d. chief
e. endocrine

20. Intrinsic factor is needed for absorption of?
a. Vitamin C
b. B12
c. A
d. E
e. B6

21. Food in the stomach is called?
a. bolus
b. chyme
c. feces
d. flatus
e. deglutition

22. Acid escaping the stomach and entering the esophagus is called?
a. deglutition

b. flatus
c. heart burn
d. bolus
e. none of the above

23. The primary function of the small intestine is?
a. water absorption
b. nutrient absorption
c. vitamin production
d. elimination
e. all of the above

24. Which is larger in diameter?
a. small intestine
b. large intestine

25. Which is longer?
a. small intestine
b. large intestine

26. The first part of the small intestine is?
a. ileum
b. jejunum
c. duodenum
d. colon
e. cecum

27. The liver, gallbladder and pancreas all release their materials into the?
a. ileum
b. jejunum
c. duodenum
d. colon
e. cecum

28. The small intestine contains a large number of?
a. microvilli
b. cilia

c. flagella
d. fundus
e. serosa

29. The liver has how many lobes?
a. 1
b. 2
c. 3
d. 4
e. 5

30. The hepatic portal vein transports materials between what two areas?
a. liver and gallbladder
b. GI tract and pancreas
c. pancreas and gallbladder
d. GI tract and liver
e. all of the above

31. The liver is divided into six sided structures called?
a. triads
b. lobules
c. cords
d. hepatocytes
e. sinusoids

32. Which is not a function of the liver?
a. bile production
b. detoxification
c. storage
d. synthesizing nutrients
e. storing bile

33. The gallbladder functions to?
a. make bile
b. store bile
c. detoxification
d. digest carbohydrates

c. store calcium

34. Bile is good for emulsifying (breaking down)?
a. carbs
b. DNA
c. proteins
d. lipids
e. all of the above

35. Most of our digestive enzymes come from the?
a. liver
b. pancreas
c. gallbladder
d. stomach
e. intestines

36. A lipase will always break down?
a. carbs
b. DNA
c. proteins
d. lipids
e. all of the above

37. Amylases will always break down?
a. carbs
b. DNA
c. proteins
d. lipids
e. all of the above

38. The small intestine and large intestine are separated by the?
a. pyloric sphincter
b. lower esophageal sphincter
c. upper esophageal sphincter
d. major duodenal papilla
e. ileocecal valve

39. Food material in the large intestine is called?
a. bolus
b. chyme
c. feces
d. flatus
e. deglutition

40. The primary function of the large intestine is?
a. water absorption
b. nutrient absorption
c. vitamin production
d. elimination
e. all of the above

41. The longest part of the large intestine is the?
a. cecum
b. colon
c. rectum
d. haustra
e. teniae coli

42. The ascending colon rises up to meet what organ?
a. pancreas
b. liver
c. gallbladder
d. spleen
e. stomach

43. Bacteria in the large intestine produce what vitamin?
a. Vitamin C
b. B12
c. A
d. K
e. B6

44. Gas in the large intestine is also called?
a. chyme
b. flatus
c. feces

d. all of the above
e. none of the above

45. If part of the GI tract passes through the diaphragm, a person has?
a. crohn's disease
b. colitis
c. hiatal hernia
d. entiritis
e. inflammatory bowel disease

Ch 7 – Answers to multiple choice questions

1. C
2. A
3. A
4. D
5. C
6. C
7. C
8. E
9. D
10. B
11. A
12. E
13. A
14. C
15. B
16. D
17. B
18. A
19. C
20. B
21. B
22. C
23. B
24. B
25. A
26. C
27. C
28. A
29. D
30. D
31. B
32. E
33. B
34. D
35. B
36. D
37. A

38. E
39. C
40. A
41. B
42. B
43. D
44. B
45. C

URINARY SYSTEM

ANATOMY OF THE KIDNEY

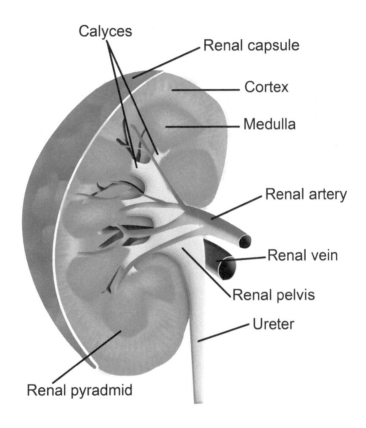

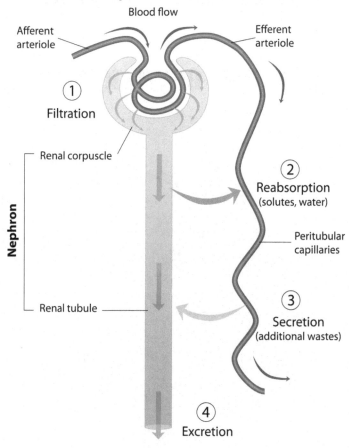

CHAPTER 8
URINARY SYSTEM

The urinary system includes the kidneys, ureters, urinary bladder and urethra. The ureters, bladder and urethra are simple in anatomy and physiology, in comparison to the kidneys. The kidneys are the number one waste removal organs in the body. They do provide other vital functions, but without their waste removal, we wouldn't live long.

URINARY SYSTEM FUNCTIONS

1. Waste removal. The kidneys are constantly filtering our blood and more specifically they are filtering the plasma. The plasma is where the waste products are located and through the blood is how the wastes are delivered to the kidneys. As parts of the plasma are removed from the body, it will become urine. Urine will primarily consist of water, wastes and excess materials.

2. Regulating blood pressure. You may not think of the kidneys when it comes to regulating blood pressure, but the kidneys are vital to blood pressure. The blood is composed almost entirely of red blood cells and water. The quantity of these two materials is what largely determines blood pressure. The kidneys have control over red blood cell production through the hormone erythropoietin. The kidneys also have control over the volume of water in the blood. Since the kidneys control blood volume, they control blood pressure. Don't forget that blood volume and blood pressure always go together.

3. Regulating red blood cell production. The kidneys have control over red blood cell production through the hormone erythropoietin.

4. Regulation of solutes. The kidneys can regulate the quantity of many solutes in the blood, especially ions. Since they regulate ion balance, they can influence all ion functions in the body.

KIDNEYS

The kidneys lie posterior to the peritoneum and are protected by the ribs on the posterior side of the two upper quadrants. Damage to the kidneys could result in a large amount of blood being lost.

A hilum is seen on the medial surface of the kidneys and just like other organs a hilum is where structures enter and leave an organ. The renal artery, renal vein and renal pelvis can be seen at the hilum. The renal arteries deliver around 20-25% of all blood to the kidneys but the sympathetic nervous system can vary this amount greatly.

The kidneys are surrounded by a thin layer called the renal capsule. This connective tissue encloses and separates the kidneys from any surrounding tissues. The inside of the kidneys are composed of two regions, the outer cortex and the inner medulla. The outer cortex is a vascular region and is composed of individual columns, which project between the pyramids of the medulla. The inner medulla is composed of separate pyramids. These pyramids are actually column shaped and the tip of each column projects towards the medial side of the kidneys. The tip of each pyramid is called a renal papilla. Think of these tips as the tips of a cone. At the tips of the renal pyramids are collecting spaces called calyces. A minor calyces is the first and smallest collecting space at the tip of a renal pyramid. This minor calyx leads to a major calyx, which is a larger collecting space. This leads to the renal pelvis and then to a ureter.

If you want to understand the kidneys, you must understand the individual functional units of the kidneys called the nephrons. The nephrons are tubes made of epithelial cells and the kidneys contain around 2.5 million nephrons. The epithelial cells will be simple cuboidal in some places and simple squamous in others. The nephron is divided into 4 main regions: Bowman's capsule, proximal tubule, loop of Henle, and distal tubule.

The Bowman's capsule is a collecting space for plasma as the fluid leaves a capillary called a glomerulus. The glomerulus is a tangled, fenestrated capillary, receiving blood from an afferent arteriole and emptying blood through an efferent arteriole. Near the afferent arteriole is a group of cells called juxtaglomerular cells. These cells come in contact with part of the nephron containing macula densa cells. The juxtaglomerular cells and the macula densa cells will release renin. Renin plays an important role in the release of aldosterone. Fenestrated means that the capillary has many tiny holes in it. When blood is pushed through the glomerulus, it must look like a soaker hose in someone's yard. When a soaker hose is pressurized, the water sprays out in all directions through hundreds of tiny holes. The glomerulus allows plasma to leave the blood and enter the collecting space around it, called the Bowman's capsule. As soon as plasma leaves the glomerulus and enters the Bowman's capsule the fluid is then known as filtrate. The Bowman's capsule plus the glomerulus is called a renal corpuscle. The Bowman's capsule is all about filtration and filtration of plasma from the glomerulus is where urine formation begins.

The glomerulus (capillary) is surrounded by a layer of epithelial cells called podocytes. Podocytes are wrapped around the capillaries to help form a mechanical filter. These cells will grow finger like extensions around the glomeruli and leave tiny spaces in between them called filtration slits. These tiny spaces are far too small for a cell to fit through, so all of the cells of the blood are held back. This is why we shouldn't see blood in the urine. The spaces are large enough that water and tiny dissolved particles like ions can move though them. This barrier of podocytes works like the filter in a coffee pot. The filter holds back the large particles, but it will allow water and dissolved particles to move through it.

After the filtration in the Bowman's capsule the remaining parts of the nephron are primarily about reabsorption. The glomerulus allowed plasma to escape the blood and now that escaped plasma is called filtrate. The remaining parts of the nephron are going to put 99% of the filtrate back into the blood by reabsorbing the nutrients back into the interstitial spaces, where a second capillary called the peritubular capillary will return it to the blood. We want to reabsorb 99% of the filtrate because only about

1% of our plasma is waste and excess materials. So when you look at the function of a nephron, it's not so much about secretion, but it is more about reabsorption. The primary job of the kidneys is not to pump wastes out of the body, their primary job is to reabsorb the materials we need.

The second part of the nephron is the proximal tubule, sometimes called the proximal convoluted tubule. After the Bowman's capsule has collected the filtrate the filtrate flows to this proximal tubule. The proximal tubule has a large number of microvilli, more than any other part of the nephron. Because of this large number of microvilli, more reabsorption of filtrate occurs here than in any other nephron part. Around 65% of all filtrate is reabsorbed in the proximal tubule.

The third part of the nephron is the loop of Henle. The loop of Henle has a descending part, which extends deep into the medulla and an ascending part which rises back up out of the medulla. On the descending part of the loop, water is pulled out of the nephron. The water is pulled out of the loop, because the deeper you go into the medulla, the high the concentration of solutes gets. Water always moves towards the solutes, so as the loop descends into the solutes, the water is pulled out. This will give an additional 15% reabsorption of materials from the nephron. Some solutes will also move into the nephron on its descent, because the solutes are diffusing from a high to low concentration.

As the loop ascends back up out of the medulla, it moves back up into lower concentrations of solutes. As this happens the solutes which just entered the nephron on the way down, come right back out of the nephron on the way up. So the net movement of solutes into and out of the entire loop is zero, but we did pull 15% of the water out of the nephron on the way down. Water channels called aquaporins are closed on the ascending part of the loop, so this prevents water from moving back into the nephron. The goal is to get the water and materials out, not into the nephron.

There are two types of nephrons and they are different because of the length of the loop of Henle. The most common types of nephrons are the cortical nephrons. These nephrons have a short loop and are not as good at reabsorption as the others. The less common type of nephron is the juxtamedullary nephron. These nephrons have a longer loop of Henle. A longer loop of Henle will extend deeper into the medulla, meaning it descends into higher solute concentrations. The higher the concentrations, the stronger the pull of water out of the loop. So more water is reabsorbed from the juxtamedullary nephrons than the cortical nephrons.

The fourth part of the nephron is the distal tubule. The distal tubules will reabsorb most of the remaining filtrate and then take the small remaining percent to a collecting duct. The distal tubule and collecting duct will usually reabsorb around 19% of the filtrate, but hormones will influence the reabsorption in the distal tubule and collecting duct. Hormones such as ADH, aldosterone and ANH work on these parts of the nephron.

If we look at our parts of the nephron, notice the reabsorption.

> Bowman's capsule = 0 reabsorption (filtration here)
> Proximal tubule = 65% reabsorption
> Loop of Henle = 15% reabsorption
> Distal tubule = 19% reabsorption

This gives a total of 99% reabsorption, meaning most all of the filtrate is moved back into the body. Only about 1% of all filtrate exits the body as what we call urine. Notice that only a tiny amount of the filtrate is lost. Most all of the filtrate is valuable material, so we must hold on to most of it.

URINE FORMATION

Blood flow to the kidneys varies greatly. The sympathetic nervous system will be constricting renal arteries when we are physically active and dilating them when we aren't. On average around 20-25% of our blood is moving to our kidneys. That is a lot of blood. If we consider a cardiac output of 5L per minute, we will move about 1.2L of blood per minute to the kidneys. Of this renal blood flow, 55% of that blood is plasma. So we are moving about .66L of plasma through the kidneys every minute. Of this plasma moving through the kidneys only about 19% will escape through the tiny holes in the glomeruli. This will give us .125L of filtrate per minute. Of this filtrate, only about 1% of it is lost, as what we call urine. The average person will produce about 1-1.5 liters of urine per day, but of course many factors affect urine output.

Urine formation is a combination of filtration, reabsorption and a smaller amount of secretion. To understand the reabsorption, we must understand basic processes, which move materials through a cell membrane. Make sure you understand the following or you won't understand the kidneys.

MOVEMENT PROCESSES

Diffusion – the movement of anything from high to low concentrations. This happens due to the random motion of atoms, so no energy is needed for this to occur.

Facilitated diffusion – This is similar to regular diffusion, but a helper (facilitator) is needed. Some materials will diffuse through a cell membrane only if a particular channel is present for it to move through. The cell can build these diffusion channels when needed.

Active transport – This involves the expenditure of ATP (energy) to pump something against diffusion. This will create high concentrations of materials. The most common example of active transport in the body is the "sodium potassium exchange pump." This pump is the driving force behind everything in the kidneys.

Cotransport (symport) – Cotransport is the movement of two materials in the same direction. This process will only work if a concentration gradient exists. As materials diffuse they will sometimes pull other materials with them. If the two move in the same direction, you have cotransport.

Countertransport (antiport) – Countertransport is the movement of materials in the opposite direction. This process will only work if a concentration gradient exists. As materials diffuse they will sometimes move something else at the same time. If the two materials move in the opposite direction, you have countertransport.

Osmosis – Osmosis is the diffusion of water. Remember that water always follows the solutes. Where the solutes go, water will follow.

REABSORPTION IN THE PROXIMAL TUBULE

In the proximal tubule you will find a simple cuboidal cell layer. On one side of the cell is the filtrate and on the other side is the interstitial space (body side). Now remember our goal is to move materials from the interstitial side, back to the body side. This all begins with the sodium potassium exchange pump on the interstitial side of the cell. This pump is pumping sodium out of the cell to create a concentration gradient. As the sodium is pumped out on the interstitial side, it can't come back in on that side, but it can come back in on the other side. So as sodium comes back in on the filtrate side, it moves through cotransporters. These cotransporters allow sodium to diffuse back in and this movement of sodium pulls other materials with it in the same direction. This will pull useful materials like sugar, amino acids, ions and other nutrients into the cell. As these solutes are pulled into the cell, water will follow it through special channels called aquaporins. This will allow water to follow the solutes into the cell. Now we have created a high concentration of nutrients inside of the cell. On the interstitial side the cell has many facilitated diffusers available for the nutrients. So the nutrients move from the inside of the cell, out the interstitial side. This is moving them back into the body. Once again the water will follow the solutes. So look what happened, the cell pumped sodium out on one side, knowing it would come back in on the opposite side. As the sodium came back in, it pulled everything else with it. This moved 65% of all filtrate from the filtrate side, back into the body. If the sodium potassium pump ever stops, then the reabsorption is lost.

Notice what is not being reabsorbed are the wastes. The wastes are being left inside the nephron, because whatever is left in the nephron is going out of the body, as what we call urine. We want to leave the wastes, excess materials and a small amount of water in the nephrons.

REABSORPTION IN THE LOOP OF HENLE

In the thin (squamous) part of the descending loop, we see water being pulled out of the nephron. The water is being pulled out because as the loop travels deep into the medulla, the concentration of solutes increases. These solutes are pulling the water out of the nephron. At the same time the solutes will also diffuse into the loop, because they are moving from a high to low concentration area. By the time we have reached the tip of the loop, we have achieved 80% total reabsorption. 65% from the proximal tubule plus 15% from the descending loop, for a total of 80%.

On the way back up the loop the water channels are closed, so the water can't move back in. Remember we are trying to move materials out of the nephron, not into it. Since the loop picked up solutes on the way down, it will lose the solutes on the way back up. As we move up the medulla, the solute concentrations drop. So the solutes will move out of the loop towards the lower solute concentrations. This gives a net solute movement of zero.

Following the loop of Henle is a capillary called the peritubular capillary and specifically a part of it called the vasa recta. This capillary has blood moving through it, in the opposite direction of the filtrate in the loop. So as the filtrate in the descending loop is traveling deeper into the medulla, the blood is traveling out of the medulla. As the filtrate in the ascending loop is traveling up out of the medulla, the blood in the capillary is traveling down deeper. This is called a counter current system, meaning we have two fluids moving in opposite directions. As the loop loses materials the blood picks them up. Remember that everything in the nephron came from the blood and our objective is to get 99% of it back into the blood. Capillaries follow the nephron and will pick up what is being reabsorbed from the nephrons. This counter current system and the solute concentrations in the medulla must exist if we are to reabsorb filtrate back into the blood.

REABSORPTION IN THE DISTAL TUBULE

The distal tubule is once again thick (cuboidal) cells. In the distal tubule, we once again see the sodium potassium pump working on the interstitial side of the cell. As before it pumps sodium out on the interstitial side and allows it to come back in on the filtrate side. As the sodium moves back in on the filtrate side, cotransporters will move other materials at the same time. This will give us more reabsorption in the distal tubule. As before water will follow these solutes, but this water movement is being controlled by hormones like ADH, ANH and aldosterone. Usually we will get an additional 19% reabsorption in the distal tubule. If we take the 65% from the proximal tubule, the 15% from the loop and the 19% here, this gives us 99% reabsorption. We will lose only about 1% of the filtrate. This filtrate being lost will be high in wastes, but will also contain water and excess materials.

The cells of the nephrons also secrete materials into the center of the nephron. The amount of secretion is low in comparison to the reabsorption, but the secretions will add to the materials becoming urine.

HORMONES WORKING ON THE DISTAL TUBULES AND COLLECTING DUCTS

ADH (antidiuretic hormone) has the strongest effect on the nephrons, when it comes to water reabsorption. Remember that diuretics cause water loss, ADH causes water reabsorption. Water balance is always directly tied into blood pressure. Water balance and blood pressure always go hand and hand. ADH will work on the distal tubules and collecting ducts by increasing sodium reabsorption. If we pull more sodium back into the body, water will follow. The ADH will also increase the production of water channels called aquaporins. So if we reabsorb more sodium and open water channels, the water will be pulled back into the body. As we pull water out of the nephron, we are decreasing urine output. We reabsorb water to help increase blood pressure.

Aldosterone works in a similar way to ADH. It will also increase the reabsorption of sodium and water. In addition aldosterone will increase the excretion of hydrogen and potassium. Aldosterone will increase the production of hydrogen and potassium antiporters. Again the sodium potassium exchange pump will create a concentration gradient for sodium. As sodium comes back into the cell, it will move hydrogen and potassium out of the cell and into the nephron. If hydrogen and potassium are being moved into the nephron, this is moving them into what will become urine. This will move them out of the body.

ANH (atrial natriuretic hormone) comes from the right atrium of the heart and it works opposite of ADH. ANH works to slow the reabsorption of sodium in the nephrons. If sodium is not being reabsorbed, then water is not being reabsorbed. If sodium stays in the nephron, this will increase urine output, which will decrease blood pressure. ANH is released when blood pressure is high. High blood pressure will stretch the right atrium. The stretching of the right atrium causes the release of ANH. ANH will cause water loss and drop the pressure back down. Remember water balance is all about blood pressure.

Renin is a hormone of the kidneys. Renin is secreted from the juxtaglomerular cells and macula densa cells. Renin through the actions of proteins and enzymes will regulate the release of aldosterone. The renin will act on a liver protein (angiotensinogen) and convert it to angiotensin I. An enzyme will convert angiotensin I to angiotensin II and this will increase aldosterone secretion.

URETERS

The ureters are hollow muscular tubes, leading from the kidneys to the urinary bladder. These tubes will use peristaltic waves to move the urine to the bladder.

URINARY BLADDER

The urinary bladder is a hollow muscular storage container. As the ureters move the urine from the kidneys to the bladder, the smooth muscle of the bladder stretches. The stretching of this muscle (detrusor muscle) is consciously detected and tells us when we need to empty the bladder. This sensation is the micturition reflex.

The bladder is lined with transitional epithelial tissue. This special type of epithelial tissue will stretch when the bladder fills. A triangular region called the trigone is located between the entry point of the two ureters and the exit point of the urethra.

URETHRA

The urethra is the exit tube for urine out of the body. The urethra is much shorter in females and this is why females are much more prone to urinary tract infections.

FILTRATE FLOW

If you look at all of the structures encountered by filtrate, you will see the following:

1. Bowman's capsule
2. proximal tubule
3. loop of Henle

4. distal tubule
5. collecting duct
6. minor calyx
7. major calyx
8. renal pelvis
9. ureters
10. urinary bladder
11. urethra

DIABETES

One type of diabetes you may not have heard of is diabetes insipidus. This diabetes involves a problem with the hormone ADH. There are several forms of the disease and all involve ADH. Consider what you would see if a person wasn't producing ADH. Without ADH the kidneys wouldn't be reabsorbing adequate amounts of sodium and water. A large volume of water would be lost and a person would have problems with dehydration, blood viscosity and blood pressure. This form of diabetes is rare and you may never know anyone with it.

What is much more common is diabetes mellitus. This is where you hear about the type I and II.

Type I diabetes mellitus is also called juvenile diabetes and accounts for about 3% of all diabetes cases. Type I is an autoimmune disease, meaning that the immune system has turned against the body and destroyed something it shouldn't. If the immune system destroys the insulin producing cells of the pancreas, type I diabetes is the result. Without the insulin producing cells, the body won't have any insulin. Without insulin most cells won't absorb sugar and amino acids. The cells of the body will starve and the nephrons will be flooded with high sugar levels. As the nephrons are unable to reabsorb this large amount of sugar, the sugar will escape in the urine. As the sugar escapes, water will escape with it. So high blood sugar levels and high urine output are common with diabetes.

Type II diabetes is also called adult onset diabetes and

accounts for about 97% of all diabetes cases. Type II is caused by people eating far too many high calorie foods and not getting enough exercise. If a person is consuming a large amount of calories and not exercising, their blood sugar levels will be chronically (always) high. When blood sugar is always high, insulin levels will always be high. When insulin levels stay high over years and decades, the body becomes tolerant to it. This means the body starts to ignore the insulin. If the body ignores insulin, the blood sugar levels stay high and many problems arise.

Most of the problems people have with diabetes mellitus, will be caused by circulation problems. They will have so many circulation problems, because their poor diet caused this problem. The fats will clog up the arteries and at the same time the chronic high blood sugar problems will do the same. As circulation is lost, this affects all areas of the body. The more vascular areas like the brain, heart, kidneys, retina, etc. are often hit hardest. These organs will slowly lose circulation and tissue will die. Amputations are also common and the further you are from the heart, the more common they will be.

Chapter 8 – Study Questions

1. Which is not a function of the urinary system?
a. waste removal
b. regulating blood pressure
c. regulating white blood cell synthesis
d. regulating solute concentrations
e. regulating water balance

2. Structures enter and leave the kidneys at a region called?
a. trigone
b. ureters
c. hilum
d. capsule
e. sinus

3. When the sympathetic nervous system is activated, blood flow to the kidneys will?
a. increase
b. decrease
c. no change

4. The kidneys are surrounded by a think connective tissue layer called the?
a. trigone
b. ureters
c. hilum
d. capsule
e. sinus

5. The outer region of the kidneys is called the?
a. capsule
b. cortex
c. hilum
d. medulla
e. trigone

6. The inner region of the kidneys is called the?
a. capsule

b. cortex
c. hilum
d. medulla
e. trigone

7. The medulla is separated into regions called?
a. cortex
b. medulla
c. capsules
d. columns
e. pyramids

8. The tip of each pyramid is called?
a. papilla
b. ducts
c. trigone
d. cortex
e. none of the above

9. At the tip of each pyramid is a collecting space called?
a. renal pelvis
b. ureter
c. minor calyx
d. major calyx
e. urethra

10. Plasma leaves the cardiovascular system and enters the nephron, when the plasma leaves the?
a. glomerulus
b. Bowman's capsule
c. loop of Henle
d. distal tubule
e. proximal tubule

11. When plasma leaves the blood and enters the nephron, it is then called?
a. interstitial fluid
b. filtrate

c. osmotic fluid
d. podocytes
e. calyx

12. The functional units of the kidneys are?
a. ureters
b. nephrons
c. podocytes
d. loops of Henle
e. pyramids

13. The filtrate collects into a space called?
a. glomerulus
b. Bowman's capsule
c. loop of Henle
d. distal tubule
e. proximal tubule

14. The glomeruli are surrounded by cells called?
a. ureters
b. nephrons
c. podocytes
d. loops of Henle
e. pyramids

15. The podocytes work as a?
a. fluid reservoir
b. mechanical filter
c. blood regulator
d. all of the above
e. none of the above

16. The Bowman's capsule is responsible for how much reabsorption from the nephrons?
a. 0%
b. 65%
c. 15%
d. 19%

e. 80%

17. What process occurs within the Bowman's capsule?
a. reabsorption
b. filtration
c. active transport
d. cotransport
e. countertransport

18. How much of the filtrate is usually reabsorbed?
a. 99%
b. 65%
c. 15%
d. 19%
e. 80%

19. Fluid leaving the Bowman's capsule will enter what?
a. proximal tubule
b. loop of Henle
c. distal tubule
d. collecting duct
e. minor calyx

20. The proximal tubule reabsorbs how much filtrate?
a. 99%
b. 65%
c. 15%
d. 19%
e. 80%

21. The loop of Henle reabsorbs how much filtrate?
a. 99%
b. 65%
c. 15%
d. 19%
e. 80%

22. The distal tubule reabsorbs how much filtrate?

a. 99%
b. 65%
c. 15%
d. 19%
e. 80%

23. What percent of the filtrate will become urine?
a. 99%
b. 5%
c. 15%
d. 19%
e. 1%

24. The movement of materials from a high to low concentration?
a. active transport
b. cotransport
c. countertransport
d. osmosis
e. diffusion

25. Some materials will diffuse through the plasma membrane only if a channel exists to assist it. This is?
a. active transport
b. cotransport
c. countertransport
d. facilitated diffusion
e. diffusion

26. The expenditure of ATP to move materials against diffusion?
a. active transport
b. cotransport
c. countertransport
d. osmosis
e. diffusion

27. The movement of two materials in the same direction is?
a. active transport
b. cotransport

c. countertransport
d. osmosis
e. diffusion

28. The movement of two materials in the opposite direction is?
a. active transport
b. cotransport
c. countertransport
d. osmosis
e. diffusion

29. The sodium potassium exchange pump is found on which side of the proximal tubule cells?
a. interstitial side
b. filtrate side

30. Nutrients are moved from the filtrate and into the cell by what process?
a. active transport
b. cotransport
c. countertransport
d. osmosis
e. diffusion

31. Water is reabsorbed from the loop of Henle where?
a. the descending portion
b. the ascending portion

32. The water reabsorbed from the loop of Henle is returned to the cardiovascular system by the?
a. vasa recta
b. afferent arteriole
c. renal artery
d. collecting duct
e. minor calyx

33. ADH will work to do what to blood pressure?
a. increase

b. decrease
c. no effect

34. ANH will work to do what to blood pressure?
a. increase
b. decrease
c. no effect

35. If a person has acidosis, you would expect aldosterone secretion to do what in response?
a. increase
b. decrease
c. no effect

36. What will transport urine from the kidneys to the bladder?
a. urethra
b. ureters
c. trigone
d. afferent arteriole
e. all of the above

37. What will transport urine out of the body?
a. urethra
b. ureters
c. trigone
d. afferent arteriole
e. all of the above

38. Diabetes insipidus involves a problem with what hormone?
a. ANH
b. ADH
c. aldosterone
d. insulin
e. renin

39. Diabetes mellitus involves a problem with what hormone?
a. ANH
b. ADH

c. aldosterone
d. insulin
e. renin

40. People with type I diabetes mellitus have?
a. high insulin production
b. low blood sugar levels
c. no insulin in the body
d. low ADH levels
e. a tolerance to insulin

41. People with type II diabetes mellitus have?
a. low insulin production
b. low blood sugar levels
c. no insulin in the body
d. low ADH levels
e. a tolerance to insulin

42. What causes type I diabetes mellitus?
a. bad diet and lack of exercise
b. too much insulin
c. autoimmune disease
d. low ADH problems
e. none of the above

43. What causes type II diabetes mellitus?
a. bad diet and lack of exercise
b. too much insulin
c. autoimmune disease
d. low ADH problems
e. none of the above

Answers to multiple choice questions.
1. C
2. C
3. B
4. D
5. B
6. D
7. E
8. A
9. C
10. A
11. B
12. B
13. B
14. C
15. B
16. A
17. B
18. A
19. A
20. B
21. C
22. D
23. E
24. E
25. D
26. A
27. B
28. C
29. A
30. B
31. A
32. A
33. A
34. B
35. A
36. B
37. A

38. B
39. D
40. C
41. E
42. C
43. A

WATER, ELECTROLYTES AND ACIDS

Chapter 9

The quantity of water in our body varies greatly. Factors such as gender, age and fat content affect how much water we retain. Males will generally have a higher water content. The average adult male will have about 60% of his body composed of water, where a female is about 50%. Much of this difference is due to the fact that males carry more muscle mass and muscle contains a large quantity of blood.

Individuals with more fats in the body will have a lower water content. Adipose tissue doesn't contain water, so the more fats a person has, the less water they possess. As we get older we lose muscle and gain adipose tissue, so as we get older we have less water in the body.

Infants have the highest water content, around 75%, but this changes rapidly as they grow.

Fluids are largely found in two places, the extracellular environment (outside cells) and the intracellular environment (inside cells). The intracellular environment is where we find the majority of our water, around 35-40% for adults with females always being the lower number. Living cells always have water inside of them. All of the chemical reactions needed to sustain life, occur in water. The extracellular environment includes many spaces containing water. The plasma, interstitial spaces, joints, eyes, etc. contain water.

Water volume is primarily maintained by the kidneys. We have seen several hormones which target the kidneys and affect water balance. Water balance is adjusted primarily to adjust blood pressure, as water balance and blood pressure always go together.

We gain water primarily through the oral cavity, meaning we drink it. The hypothalamus has osmoreceptors which are always monitoring the solute concentrations (osmolality) of the blood. If the osmoreceptors determine that our blood has too many particles in solution, this means we don't have enough water in it. The hypothalamus will tell us we are thirsty and when we drink fluids this will thin our blood back out. ADH will be released at the same time. Think of these osmoreceptors as viscosity receptors. If our blood viscosity is too high, meaning it's too thick, then we need more water in it. That is when the hypothalamus tells us we are thirsty.

In addition to drinking water we also gain water through anabolism. As our cells are building materials, hydrogen and oxygen are often released. These atoms will combine to make water and will be used in the body.

We lose water through the urinary, respiratory, integumentary and digestive systems. The kidneys are the primary regulator of water through urine output. Every time we inhale we add moisture to air and then when we exhale that water is lost. When we are hot a large amount of water can be lost through sweat glands and a small amount of water is lost in feces. Through all of the ways we take water in and lose water, total water balance remains the same. What we take in, is generally what we lose.

HORMONES AFFECTING WATER BALANCE

Antidiuretic hormone is the primary hormone associated with water balance. The hypothalamus uses osmoreceptors to monitor blood thickness and will regulate ADH release to balance blood viscosity. When our osmolality is too high, our blood is too thick. When our blood is too thick, we need more water in it. This is when ADH is released. If we are low on water, we don't want to lose more through the kidneys. So water is held by the reabsorption of sodium. When the kidneys retain sodium the water will follow and urine output drops.

The hypothalamus also uses baroreceptors to monitor blood pressure and regulate water balance. Baroreceptors monitor pressure by monitoring the stretching of the aorta and the internal carotids. If the baroreceptors are only being stretched a little, they will send action potentials to the brain infrequently. A low frequency is seen as low blood pressure. When our blood pressure is low, we want to hold on to water. ADH will be released at this time, telling the kidneys to hold water. When blood pressure is high, we want to lose water. Less ADH is released at this time and urine output increases. Water balance and blood pressure always go together.

Aldosterone will also cause sodium and water retention in the kidneys. Water retention is desired if blood pressure is low. When blood pressure is high, we want to lose water.

Atrial natriuretic hormone (ANH) works opposite of ADH and aldosterone. ANH will cause water to be lost, not retained. When blood pressure is high, we see more stretching of the right atrium. As the right atrium is stretched, more ANH is released. ANH will cause the kidneys to release more sodium, resulting in the loss of water. As we lose water, blood pressure will drop.

SODIUM REGULATION

Sodium regulation is primarily regulated by ADH and aldosterone. These hormones will cause sodium retention, while ANH will cause sodium excretion. Since these hormones target the kidneys, the kidneys are the primary organs associated with sodium balance.

Sodium is the dominant cation (positive ion) of the body. Sodium is by far what moves water, so it is vital to water balance. Sodium is also the dominant extracellular cation when it comes to membrane potentials. Membrane potentials always refer to the charges found on a cell when it is at rest. The outside positive charge of the cell, comes predominantly from the high concentration of sodium in the extracellular environment.

A large amount of sodium can be lost through the integumentary system, when we sweat. When the sweat glands want to move water, they move sodium and the water will follow. The hypothalamus is the temperature control center of the body, so it is the hypothalamus regulating the sweat glands.

Hyponatremia is a low blood sodium level and hypernatremia is an elevated blood sodium level. Too little or too much of anything in the body will cause problems. Since sodium largely works to move water, water will be unbalanced if sodium is unbalanced. When water is imbalanced then blood pressure will be too high or too low.

Hyponatremia can be caused by excessive sweating or any time when large volumes of water are lost. Excessive urination, vomiting, etc. will result in low sodium levels.

Hypernatremia is often caused by too much sodium taken in with food. Hypersecretion of aldosterone or ADH could also be a cause. These hormones cause retention of sodium, so if too much is secreted, then sodium levels will be high.

If sodium becomes imbalanced in the body, expect to see water problems, blood pressure problems and problems with the nervous system and muscular system.

POTASSIUM REGULATION

Potassium is largely under the regulation of aldosterone. Aldosterone affects sodium, hydrogen and potassium balance in the kidneys. Aldosterone causes the kidneys to excrete potassium, so if potassium levels are high, more aldosterone will be released. If potassium levels are low, then less aldosterone is released.

If potassium becomes unbalanced, membrane potentials will be upset. Potassium has a large effect on the resting membrane potential. This ion is found in lower quantities than calcium and sodium, so it is easier to upset its balance. Also we don't have potassium storage sites, so that also makes it easy to upset potassium balance. If potassium becomes unbalanced, membrane potentials will be upset and the nervous system and muscular system won't work properly.

Hypokalemia is too little potassium and hyperkalemia is too much potassium.

CALCIUM REGULATION

Calcium is largely regulated by calcitonin and parathyroid hormone (PTH). Calcium is needed for more than bone strength. The hard mineral part of bone (hydroxyapatite) requires calcium, but we also need it for blood clotting, muscle contraction and many other functions.

Calcitonin will inhibit osteoclast activity. Osteoclasts break down bone, when calcium release is needed. So if calcitonin inhibits osteoclasts then calcitonin is slowing the release of calcium. We want to inhibit calcium release when blood calcium levels are adequate or high. Calcitonin is working to lower blood calcium levels.

Parathyroid hormone has several target tissues. PTH will stimulate osteoclast activity. If osteoclasts are stimulated then more calcium is released. This will help to raise blood calcium levels. PTH also targets the kidneys and causes more calcium to be reabsorbed. This will help raise blood calcium levels. PTH will target the small intestine and cause calcium absorption. This will help raise blood calcium levels. PTH will increase Vitamin D production and this will raise blood calcium levels. So PTH is working to raise blood calcium levels.

ACID BASE BALANCE

Acid base balance is primarily about hydrogen ion balance. We have several systems working to balance pH at all times, but the respiratory system is the best system for pH balance. pH must remain within the 7.35-7.45 range if we are to remain within homeostasis. The respiratory system is our best pH balancing system, because of the relationship between carbon dioxide and hydrogen ion. Recall the following chemical reaction.

$$CO_2 + H_2O \leftrightarrow H_2CO_3 \leftrightarrow H + HCO_3$$

This chemical reaction tells us two things:

1. The reaction is reversible, so whatever happens to one side happens to the other.

2. Carbon dioxide and hydrogen ion are on different sides of the equation. So whatever happens to carbon dioxide affects hydrogen and vice versa. If carbon dioxide levels increase, hydrogen ion levels increase. If one of them decreases, the other decreases.

So if we have a pH imbalance, we have a hydrogen ion imbalance. If hydrogen is out of balance then carbon dioxide must be out of balance. When pH gets below 7.35, this means there are too many hydrogen ions in the blood and the condition acidosis develops. If we have too many hydrogen, we have too much carbon dioxide. Ventilation will increase, because when we exhale, we release carbon dioxide. When carbon dioxide is taken out of the body, hydrogen ion is taken out of the body. Hydrogen is not being released with respiration, it is being removed through the shifting of the chemical reaction to the left.

If we have a pH above 7.45, this means we don't have enough hydrogen in the blood. If we don't have enough hydrogen, then we don't have enough carbon dioxide. At this time the brain would slow ventilation, to retain more carbon dioxide. If we retain carbon dioxide, we will gain more hydrogen in the blood. An increase in hydrogen levels will fix the alkalosis.

Persons with acidosis tend to have rapid ventilation, while a person with alkalosis tends to have slow ventilation.

Aldosterone will signal the kidneys to secrete hydrogen. If a person has acidosis, then aldosterone levels will rise in response. More aldosterone will increase hydrogen excretion and this will help to fix the acidosis problem.

Plasma proteins and negative ions will also work as buffers in the blood.

Chapter 9 – Study Questions

1. Who carries more water in the body?
   a. males
   b. females
   c. neither

2. Individuals carrying more fats tend to carry?
   a. same amount of water
   b. less water
   c. same amount of water

3. Who carries the most water?
   a. males
   b. females
   c. infants

4. We find more water where?
   a. intracellular environment
   b. extracellular environment
   c. equal amounts in both

5. Water balance is regulated primarily by what system?
   a. integumentary
   b. digestive
   c. respiratory
   d. urinary
   e. none of the above

6. If blood osmolality increases, this means the blood is too?
   a. thin
   b. thick
   c. neither

7. When blood osmolality is too high, what will happen to ADH secretion in response?
   a. more ADH
   b. less ADH
   c. no change

8. When baroreceptors are sending a low frequency of action potentials to the brain, what will happen to ADH release?
a. more ADH
b. less ADH
c. no change

9. ANH will cause?
a. water loss
b. water retention
c. neither

10. ADH helps to do what to blood pressure?
a. raise
b. lower
c. no change

11. ANH helps to do what to blood pressure?
a. raise
b. lower
c. no change

12. ADH and aldosterone will work to?
a. retain sodium
b. lose sodium
c. no change

13. When sodium is lost, water is?
a. retained
b. lost
c. no change

14. Potassium is largely regulated by what hormone?
a. ADH
b. ANH
c. aldosterone
d. calcitonin
e. PTH

15. If a person is secreting too much aldosterone, you would soon see potassium levels being?
a. to high
b. to low
c. no change

16. Which hormone works to lower blood calcium levels?
a. ADH
b. ANH
c. aldosterone
d. calcitonin
e. PTH

17. Which hormone works to raise blood calcium levels?
a. ADH
b. ANH
c. aldosterone
d. calcitonin
e. PTH

18. Calcitonin will do what to osteoclast activity?
a. increase
b. decrease
c. no change

19. Normal pH range is
a. 7.0-7.20
b. 7.35-7.45
c. 7.30-7.40
d. 7.45-7.55
e. none of the above

20. What system is best at regulating pH levels?
a. urinary
b. digestive
c. integumentary
d. respiratory

e. endocrine

21. A person with acidosis has?
a. too many hydrogen ions in the blood
b. too few hydrogen ions in the blood

22. A person with acidosis has?
a. too much carbon dioxide in the blood
b. too little carbon dioxide in the blood

23. If a person develops acidosis, the ventilation will respond by being?
a. rapid
b. slow

Answers to multiple choice questions.

1. A
2. B
3. C
4. A
5. D
6. B
7. A
8. A
9. A
10. A
11. B
12. A
13. B
14. C
15. B
16. D
17. E
18. B
19. B
20. D
21. A
22. A
23. A

REPRODUCTIVE SYSTEM

Male Reproductive System

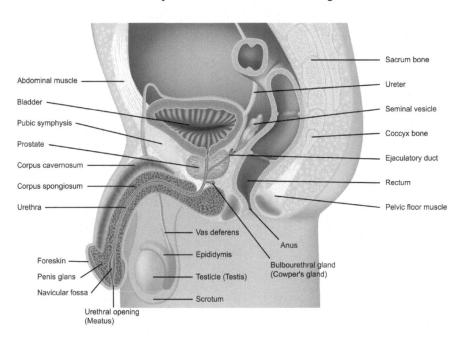

FEMALE REPRODUCTIVE SYSTEM

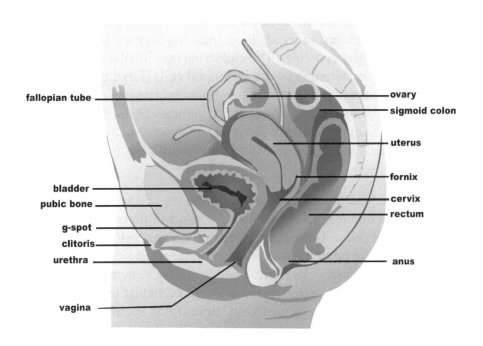

CHAPTER 10
REPRODUCTIVE SYSTEM

The reproductive system is the one system that the individual doesn't need to survive. This system doesn't keep the individual alive, it keeps the species alive. This is also the only system which shows any real difference between the two genders. Obviously the male and female vary greatly with this system. Also, the female reproductive system is more complex than the males.

MALE REPRODUCTIVE SYSTEM

The testes are the gonads of the male reproductive system. Gonads are the organs that produce the reproductive cells of the body. The gonads are a mixed gland, meaning they are part endocrine and part exocrine. The endocrine part involves the production of testosterone by the interstitial cells (Leydig cells) of the testes. The exocrine part like any other involves ducts and the testes have a complex duct system. The gametes of the male are the sperm cells. Gametes will always refer to the reproductive cells. The sperm cells are produced within the seminiferous tubules of the testes.

If we look at the inside of the testes, we will see many seminiferous tubules. In between all of the seminiferous tubules are the interstitial cells. These are the cells producing testosterone. The tubules are made up of a basement membrane and a simple columnar layer of sustentacular cells (Sertoli cells). These sustentacular cells form the blood testes barrier. This is a very particular barrier of epithelial cells, working to protect the developing sperm cells. The immature sperm cells (spermatogonia) are found around the outer area of the seminiferous tubules and as you travel deeper in, the sperm cells are more mature. A sperm cell has a head, body and tail. The head contains the genetic material along with a cap called an acrosome. The acrosome is a pack of enzymes used to penetrate a protective layer of cells surrounding the female gamete (oocyte). The body is filled with mitochondria and the tail is a flagella. A flagella is always used for propulsion. The flagella are like an outboard motor on the back of the cell and are the only place in the human body, where a flagella can be found. Sperm cells only develop properly at 95 F (35 C). This is why the testes are found outside of the body. If the testes fail to descend out of the abdominal cavity (cryptorchidism), the sperm cells will be too warm. This will prevent the sperm cells from developing properly and the male will be infertile.

The gametes develop by the process of meiosis. Meiosis is very similar to mitosis, but in meiosis you go from one cell to four, instead of one to two and the end cells have half copies of genetic material instead of complete copies.

The testes are surrounded by a tough connective tissue layer called the tunica albuginea. The tunic albuginea extends deep into the testes with extensions called septa. The septa separate the testes into compartments and the compartments are filled with seminiferous tubules.

On the external surface of the testes is the epididymis. The epididymis is the final site of sperm cell maturation. The epididymis connects to the ductus deferens (vas deferens) and this duct leads back up into the abdominal cavity. A vasectomy is the cutting of the ductus deferens. If these ducts are cut and tied, then sperm cells can't leave the body. This will make the male sterile.

The testes are found outside the body in a muscular container called the scrotum. The scrotum is largely made up of a muscle called the dartos muscle. This muscle, along with the cremaster muscle, is used to regulate the temperature of the sperm cells. If the sperm cells get to cold (below 95F) the muscle will contract, pulling them closer to the body, which will warm them. If the sperm cells get to warm (above 95F) the muscle will relax, moving them away from the body, which will cool them. The scrotum has a midline elevation called the raphe.

The testes develop high in the abdominal cavity and will descend and leave the cavity just before birth in most males. The testes pass through the peritoneum in the front of the abdominopelvic cavity. Where the testes pass they leave two weak spots in the front of the abdomen called inguinal canals. Sometimes small intestine will make its way into these canals and become constricted. This constriction could cause a death of the tissue. If something passes into the inguinal canals, this is called an inguinal hernia. These hernias often occur when a male is lifting a heavy weight. Cryptorchidism is a failure of one or more of the testes to descend.

When sperm cells leave the epididymis, they travel through the ductus deferens. The ductus deferens passes through the inguinal canals and into the pelvic cavity. After passing over the bladder they will meet the seminal vesicles. Together the two form the ejaculatory duct and pass through the prostate gland. The prostate is the only unpaired gland used in the production of semen. Unpaired meaning there is only one of them. The testes, bulbourethral glands and seminal vesicles are all paired (two of them).

The prostate gland has the beginning of the urethra, which is common to the urinary system also. As materials pass through the urethra, they will soon encounter the bulbourethral glands. After the bulbourethral glands is the penis and the spongy urethra running through it.

The penis is comprised of three columns of tissue. The corpus spongiosum is found medially with the two corpus cavernosum located laterally on both sides. These are the three columns of tissue associated with erection. Erection doesn't occur due to muscle in this tissue. Erection occurs when the blood vessels leading into these tissues dilate and the ones leading out constrict. This will cause an increase in blood pressure inside of the tissues, just like an inflated balloon. Viagra and similar drugs work as vasodilators. The vasodilation will increase blood flow to the penis, thus causing erection. Since the reproductive system is stimulated by the parasympathetic nervous system, acetylcholine causes erection naturally. The tip of the penis forms an enlarged end called the glans penis. At birth a fold of skin called the foreskin covers the glans, but is often removed by the process of circumcision.

SEMEN

Semen refers to the materials released by the male at the time of ejaculation. The semen is made by seven different glands in the male reproductive system (2 testes, 2 bulbourethral glands, 2 seminal vesicles and 1 prostate). The two testes contribute only 5% of the components of semen and of this 5%, less than 1% of it is sperm cells. The two bulbourethral glands contribute 5%. The two seminal vesicles contribute 60%, the greatest amount. The one prostate contributes 30%.

HORMONES

The hormones of the male reproductive system begin with gonadotropin releasing hormone (GnRH) of the hypothalamus. The GnRH regulates the release of luteinizing hormone (LH) and follicle stimulating hormone (FSH). The LH stimulates the production of testosterone and the FSH stimulates sperm cell production. Testosterone affects many tissues of the male body. It causes an increase in muscle mass, stimulates growth, a maturing of the reproductive system, increases red blood cell production and many other functions.

FEMALE REPRODUCTIVE SYSTEM

The ovaries are the gonads of the female reproductive system. The ovaries are a mixed gland, just like the testes of the male. The endocrine part involves the hormones estrogen and progesterone. The uterine tubes make the exocrine part. The gametes (reproductive cells) of the female are the oocytes and it is the oocytes which travel through the uterine tubes.

The ovaries are divided into two regions, cortex and medulla. The cortex is the outer region and the medulla is the inner region. The outer cortex contains the developing oocytes and the inner medulla contains blood vessels, nerves and lymphatic vessels. The ovaries are held in place by ligaments, such as the suspensory ligaments and ovarian ligaments.

The oocytes begin as immature cells called oogonia. The ovaries contain millions of these cells in embryonic development, but only about 400 will reach full maturity. The primordial follicles of the ovaries will develop into primary follicles. The primary follicles will develop into secondary follicles and a secondary will develop into a mature (Grafian) follicle. The oocytes are released from the mature follicle at the time of ovulation. After ovulation what is left of the mature follicle becomes the corpus luteum. The corpus luteum is the primary production site of progesterone. If pregnancy occurs, the corpus luteum remains active and if pregnancy doesn't occur it will degenerate into the corpus albicans. Oocytes are produced by meiosis, just like the sperm cells of the male. The sperm cell of the male joins with the oocyte of the female in a process called fertilization and forms the first cell called the zygote. Where the gametes have 23 chromosomes the zygote will have 46.

The oocyte is released into the ovarian tubes (fallopian tubes) at the time of ovulation. The ovarian tubes are filled with cilia and the cilia are what move the oocyte to the uterus. Where the ovarian tubes meet the ovaries, there is a loose connection called the fimbria. On very rare occasions the oocyte or zygote might get between the fimbria and ovary and leave the reproductive system. Sometimes a pregnancy occurs somewhere it shouldn't, this is an ectopic pregnancy. The tubes are widened near the ovaries. This wide part of the tubes is the infundibulum. The infundibulum is where fertilization will usually occur. As the tubes near the uterus, they will become narrow in a region called the isthmus.

The uterus is a hollow muscular organ made mostly of smooth muscle. The outer layer is the perimetrium. The middle layer is the myometrium and is the thick smooth muscle layer. The inner layer is the endometrium and this is the layer which is lost every month during menses. The uterus is held in place primarily by the broad ligament.

The superior part of the uterus is the wide fundus. The body is the main part of the organ and the cervix is the narrow inferior part attaching to the vagina. The vagina forms a circular region around the cervix called the fornix.

The vagina is a hollow tube of smooth muscle located between the rectum and the urethra. The vagina will extend from the labia on the outside to the cervix of the uterus. The vagina is filled with rugae (folds) like seen in the stomach. The rugae will allow for expansion when receiving the penis or at the time of child birth. A thin membrane called the hymen will partially cover the vagina until the first sexual intercourse or it can be torn during physical activity.

At the opening to the vagina is the vulva, which refers to all of the external reproductive structures surrounding the vagina. The labia minor are the thin layers on either side of the vagina and the space between them is the vestibule. Lateral to the labia minor are the labia major. The major are the larger folds surrounding the vagina. Anterior to the vagina is the urethra and anterior to the urethra is the clitoris. The clitoris becomes engorged with blood during sexual intercourse, like the penis does. Anterior to the clitoris is the mons pubis, which is a fat pad.

The menstrual cycle refers to the normal cycle of hormones and the endometrium of the uterus in nonpregnant females. Menses is separated into two distinct phases, proliferative and secretory phases. The entire cycle takes around 28 days on average, so the two phases last 14 days each. In each phase pay close attention to what is happening with each hormone, the developing follicles and the endometrium.

Menses starts with the hypothalamus and gonadotropin releasing hormone (GnRH). GnRH stimulates the release of FSH and LH. These two hormones will begin to rise in the proliferative phase and then rise sharply just before ovulation in response to rising estrogen levels. As FSH levels rise, this will stimulate the development of follicles. As the follicles mature, they secrete more estrogen and usually only one follicle reaches the mature stage. Notice how LH, FSH and estrogen all reach peak levels at ovulation.

Ovulation is the release of the oocyte from the ovaries. As LH rises it will reach high enough levels to trigger ovulation. In response to the rising estrogen levels, the endometrium of the uterus undergoes mitosis. This causes a building of the endometrium.

The secretory phase begins after ovulation has occurred. LH, FSH and estrogen all decline in this phase, where they were all rising in the first. After the oocyte has been released from the mature follicle, the remaining follicle cells are now called the corpus luteum. The corpus luteum will secrete large amounts of progesterone in this phase. So progesterone levels are much higher in the second phase than in the first. Progesterone will cause the endometrial cells to undergo hypertrophy (enlarge). So the endometrium has undergone growth through menses. Estrogen caused growth in the proliferative phase and progesterone caused growth in the secretory phase. If pregnancy doesn't occur estrogen and progesterone levels will decrease and the spiral arteries in the endometrium will constrict. At this time the endometrium will be lost and expelled from the uterus. If pregnancy does occur these two hormones increase and the endometrium will remain. The hormone human chorionic gonadotropin (HCG) is responsible for their continued production. The ending of menses is a key indicator of pregnancy.

Menopause is the ending of menses and usually occurs in the 40's or 50's.

The mammary glands are the milk producing exocrine glands of the integumentary system. The center of each breast has a raised nipple region, with a surrounding pigmented area called the areola. The breasts will grow in response to rising estrogen and progesterone levels during puberty. The breast contains a large amount of adipose tissue and this tissue largely accounts for its size and shape. The tissue is suspended by mammary ligaments, which are dense regular collagen arrangements. The mammary glands remove nutrients from the blood and concentrate the materials into the ducts. The ducts all lead to the surface of the nipple, where the materials will be released.

Chapter 10 – Study Questions

1. The gonads are the organs which do what?
a. produce blood cells
b. protect us from foreign invaders
c. regulate wastes
d. produce reproductive cells
e. regulate body temperature

2. The gonads of the male are?
a. ovaries
b. testes
c. penis
d. prostate gland
e. seminal vesicles

3. Sperm cells are produced inside of the?
a. prostate gland
b. seminiferous tubules
c. seminal vesicles
d. bulbourethral glands
e. all of the above

4. Which is not part of a sperm cell?
a. epididymis
b. head
c. body
d. tail
e. all are part

5. Sperm cells only develop properly at what temperature?
a. 90F
b. 92F
c. 95F
d. 96F
e. 98.6F

6. The condition in which the testes fail to descend?
a. vasectomy

b. impotence
c. resolution
d. cryptorchidism
e. none of the above

7. Gametes develop by the process of?
a. mitosis
b. meiosis
c. cryptorchidism
d. resolution
e. puberty

8. The sperm cells complete development where?
a. epididymis
b. testes
c. prostate gland
d. seminal vesicle
e. bulbourethral gland

9. Testosterone is produced by the?
a. spermatogonia
b. Leydig cells
c. spermatids
d. septa
e. tunica albuginea

10. The cutting of the vas deferens to cause sterilization is?
a. vasectomy
b. impotence
c. resolution
d. cryptorchidism
e. none of the above

11. The testes are found inside of the?
a. scrotum
b. inguinal canals
c. corpus spongiosum
d. corpus cavernosum

e. ampulla

12. What structure is used to regulate sperm cell temperature?
   a. epididymis
   b. ductus deferens
   c. scrotum
   d. corpus spongiosum
   e. all of the above

13. Sperm cells move from the testes to the pelvic cavity through what structures?
   a. epididymis
   b. ductus deferens
   c. scrotum
   d. corpus spongiosum
   e. all of the above

14. The majority of semen is produced by what glands?
   a. testes
   b. seminal vesicles
   c. prostate
   d. bulbourethral glands
   e. none of the above

15. The only unpaired gland is the?
   a. testes
   b. seminal vesicles
   c. prostate
   d. bulbourethral glands
   e. none of the above

16. Sperm cells make up what % of semen?
   a. 1%
   b. 5%
   c. 30%
   d. 60%
   e. 99%

17. The prostate gland makes what % of semen?
a. 1%
b. 5%
c. 30%
d. 60%
e. 99%

18. The gonads of the female reproductive system are?
a. uterus
b. uterine tubes
c. clitoris
d. vagina
e. ovaries

19. The gametes of the female reproductive system?
a. corpus luteum
b. oocytes
c. fimbria
d. isthmus
e. fundus

20. The outer region of the ovaries is the?
a. cortex
b. medulla
c. follicle
d. luteum
e. perimetrium

21. The corpus luteum secretes mostly?
a. estrogen
b. testosterone
c. progesterone
d. luteinizing hormone
e. follicle stimulating hormone

22. The gametes contain how many chromosomes?
a. 60

b. 30
c. 23
d. 46
e. 15

23. Fertilization will usually occur in the?
a. infundibulum
b. isthmus
c. endometrium
d. fundus
e. cervix

24. The uterus is composed mostly of?
a. blood
b. endometrium
c. smooth muscle
d. collagen
e. ligaments

25. This layer is lost during menses?
a. endometrium
b. myometrium
c. perimetrium
d. serosa
e. fimbria

26. The superior part of the uterus is the?
a. serosa
b. body
c. cervix
d. fundus
e. ampulla

27. The inferior part of the uterus is the?
a. serosa
b. body
c. cervix
d. fundus

e. ampulla

28. What membrane will partially cover the vagina?
a. serosa
b. body
c. cervix
d. fundus
e. hymen

29. The external reproductive structures surrounding the vagina are the?
a. serosa
b. body
c. vulva
d. fundus
e. hymen

30. The most anterior structure is the?
a. vagina
b. clitoris
c. urethra

31. The hormone responsible for ovulation is?
a. estrogen
b. progesterone
c. follicle stimulating hormone
d. luteinizing hormone
e. prolactin

32. Ovarian follicles will develop because of which hormone?
a. estrogen
b. progesterone
c. follicle stimulating hormone
d. luteinizing hormone
e. prolactin

33. Estrogen causes the endometrium to grow during which phase?
   a. proliferative
   b. secretory

34. Progesterone causes the endometrium to grow during which phase?
   a. proliferative
   b. secretory

35. Estrogen is highest during which phase?
   a. proliferative
   b. secretory

36. Progesterone is highest during which phase?
   a. proliferative
   b. secretory

Answers to multiple choice questions.

1. D
2. B
3. B
4. A
5. C
6. D
7. B
8. A
9. B
10. A
11. A
12. C
13. B
14. B
15. C
16. A
17. C
18. E
19. B
20. A
21. C
22. C
23. A
24. C
25. A
26. D
27. C
28. E
29. C
30. B
31. D
32. C
33. A
34. B
35. A
36. B

DEVELOPMENT

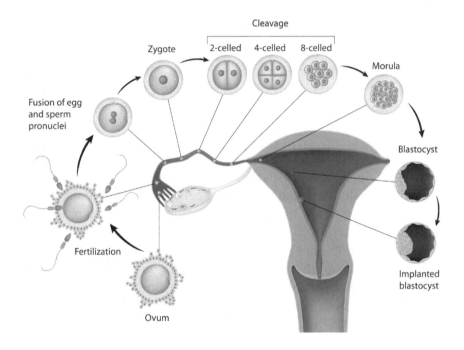

CHAPTER 11
DEVELOPMENT

The gametes of the body are the reproductive cells and these are the only cells, which have 23 chromosomes. These cells are called haploid cells and think of the "h" in haploid as standing for half, because these cells have half as many chromosomes as the other cells of the body. The other cell types of the body are called diploid cells. These cells have 46 chromosomes in their nucleus. Think of the "d" in diploid as standing for double, because these cells have twice as many chromosomes as the gametes.

The haploid cells are developed by a process called meiosis. Meiosis is very similar to mitosis, with a few differences. You will still see the same stages of mitosis; prophase, metaphase, anaphase and telophase, but there will be two divisions in meiosis, where mitosis involves one division.

With mitosis a cell will divide and then go back into interphase. Interphase involves the replication of the DNA, that way when the cell divides again, each new cell will get a complete set of DNA. With meiosis the cell skips interphase between the two divisions. If interphase is skipped, then the DNA is not duplicated, so when the cell divides a second time, the daughter cells will end up with half copies of DNA. Mitosis is where you go from one cell to two cells. Meiosis is where you go from one cell to four. This creation of sperm cells is called spermatogenesis and the creation of oocytes is called oogenesis.

FERTILIZATION

When semen is released into the vagina the sperm cells will undergo a change called capacitation, which will involve the release of enzymes. If fertilization is to occur, it must happen within 24 hours of the release of the oocyte. The oocyte lives for one day, where the sperm cells can live for several days inside the female reproductive system. When the oocyte is released the cilia in the uterine tubes will move the oocyte towards the uterus and this takes several days. Since the trip through the uterine tubes takes several days and the oocyte only lives for one day, then fertilization must take place back close to the ovaries. The sperm cells possess a flagella and this flagella will propel them up the uterus, through the uterine tubes and towards the ovaries. Somewhere near where the ovaries and uterine tubes meet is where the two gametes will fuse in a process called fertilization.

The sperm cells will be attracted to the oocytes by chemotaxis, the process by which cells follow chemical trails. If a sperm cell reaches an oocyte, the sperm cell must penetrate a barrier of cells called the corona radiate. Deep to the corona radiate is thin layer called the zona pellucida. The acromosome on the tip of the sperm cell will dissolve these outer layers and allow the sperm cell to reach the ovum. When the sperm cell reaches the plasma membrane of the oocyte, the oocyte will release calcium, followed by a release of water. This will prevent any more sperm cells from penetrating the oocyte. The genetic material of the two cells will fuse and this fusing results in the creation of the zygote, the first original cell of us all. The zygote is a diploid cell, have 46 chromosomes, 23 from the mother and 23 from the father.

The prenatal period begins with the creation of the zygote. The prenatal period can be broken down into three periods, the germinal, embryonic and fetal. The germinal period is the first two weeks of development. This period is named because of the development of the germ layers, the endoderm, ectoderm and mesoderm. The second period (embryonic) falls between weeks two and eight. During this time all of the organs of the body develop. The last period is the fetal period, weeks 8-delivery. In the fetal period all of the organs and organ systems are developing.

About 24 hours after fertilization, the zygote will start to divide. These dividing cells are called pluripotent cells, meaning these cells have the ability to change into the adult cells of the body. Soon the zygote will become a small mass of cells called a morula. The morula will continue to divide and begin to form a hollow sphere called a blastocyst.

IMPLANTATION

About 1 week after fertilization the blastocyst will implant into the endometrium of the uterus. The blastocyst contains an inner cell mass from which the embryo will develop. The inner cell mass will also form two hollow spaces called the yolk sac and the amniotic cavity. The amniotic sac will form a layer of water around the developing fetus. There is also an outer layer called the trophoblast. The trophoblast will form the placenta and the chorion. The chorion and its associated villi are blood vessels allowing exchange between the mother's blood and embryonic blood. The chorion and villi will be part of the placenta, which will allow for the swapping of materials and an anchor for the embryo.

PLACENTA

The placenta is an organ only seen during pregnancy. The placenta forms a very particular barrier between the mother's blood and the baby's blood. If everything goes as it should, the two blood supplies never mix, but many materials will cross the placental barrier. For the embryo to survive and develop nutrients must pass to the embryo from the mother and wastes must pass from the embryo to the mother. Most harmful substances can't pass the placenta, but lipid soluble drugs and some microorganisms can. HCG (human chorionic gonadotropin hormone) is secreted from the placenta and the HCG will keep estrogen and progesterone levels high. This organ will also hold the embryo and fetus in place to the uterine wall.

A layer of cells called the syncytiotrophoblast will develop from the placenta and invade the uterine wall. The syncytiotrophoblast invades the uterine wall in search of maternal blood vessels. Large pools of maternal blood will be formed and the villi will invade this area.

GERM LAYERS

In the first few weeks of development the original embryonic tissues called the germ layers will develop. These layers are the endoderm, ectoderm and mesoderm.

The endoderm (inner layer) will later develop into the deeper organs of the body. We see linings of the lungs, linings of the GI tract, a few GI tract ducts, tonsils, thyroid, thymus and parathyroids developing from this layer.

The ectoderm (outer layer) will develop into the superficial structures of the body. The epidermis, tooth enamel, cornea, lens and many parts of the nervous system develop from this layer.

The mesoderm (middle layer) will form the skeletal muscle, bones, dermis, blood vessels and many glands. Some of the mesoderm will develop into sections called somites. These somites will develop into the skull, vertebral column and skeletal muscles.

DEVELOPMENT

The face develops from five different structures: the frontonasal process, the two maxillary processes and the two mandibular processes. If these structures fail to fuse properly a cleft lip or cleft palate may result.

Skeletal muscles develop from myoblasts before birth. The number of muscle cells we are born with are all we will ever have. We are adding size to these cells as we grow.

The upper and lower limbs will begin development after 4 weeks. The proximal tissues will develop first and then progress out. All of an upper or lower limb isn't made at the same time.

There are a few changes in circulation around the time of birth. Before birth there is a hole between the two atria called the foramen ovale. When we take our first breaths changes in pressure will cause two flaps of tissue to fuse together and close off this passage. Where there was a hole (foramen ovale) there will be a depression (fossa ovalis) as an adult. Sometimes they don't fuse completely and this is the hole in the heart that newborns will sometimes have. Another circulation change is seen between the pulmonary trunk and the aorta. The ductus arteriosis connects these two structures before birth, but the small artery between them will be constricted and the ductus arteriosis will become the ligamentum arteriosis. This ligament can still be seen in adults, binding the pulmonary trunk and aorta.

PARTURITION

Parturition is just a fancy way of saying labor. Labor is divided into 3 stages.

Stage 1 – It begins with regular uterine contractions and ends when the cervix is fully dilated, meaning it is about the size of the baby's head.

Stage 2 – This stage begins with the dilated cervix and ends when the baby exits the vagina.

Stage 3 – This stage involves the expulsion of the placenta from the uterus.

If necessary the baby can be surgically cut from the uterus and abdomen in a procedure called cesarean section (C section).

Babies born premature will often have respiratory problems. It is after the 7th month of development, that the lungs begin producing surfactant. Surfactant decreases the attraction of water molecules in the lungs, so without it the lungs are more difficult to expand. The muscles of ventilation can become exhausted and the newborn could die.

TWINS

Twins come in two forms: monozygotic (identical) and dizygotic (fraternal). Identical twins come from one fertilization, meaning one sperm cell and one oocyte made one zygote. The zygote split and two individuals develop with the same DNA. Fraternal twins come from two fertilizations. One sperm fertilized one oocyte and a different sperm fertilized a different oocyte. The individuals develop and they are no more genetically related than two siblings born years apart. Fraternals can be of different genders; where identical will be the same gender.

Chapter 11 – Study Questions

1. The reproductive cells of the body are called?
a. gonads
b. gametes
c. diploid cells
d. embryo cells
e. none of the above

2. Haploid cells have how many chromosomes?
a. 16
b. 23
c. 32
d. 36
e. 46

3. The reproductive cells of the body are produced by what process?
a. meiosis
b. mitosis
c. fertilization
d. capacitation
e. none of the above

4. In meiosis what stage is skipped in between divisions?
a. interphase
b. prophase
c. metaphase
d. anaphase
e. telophase

5. After sperm cells are released they undergo a change called?
a. meiosis
b. mitosis
c. fertilization
d. capacitation
e. chemotaxis

6. After ovulation the oocyte must be fertilized within how many hours?
   a. 1
   b. 2
   c. 6
   d. 12
   e. 24

7. The oocyte is moved through the uterine tubes by the?
   a. microvilli
   b. flagella
   c. cilia
   d. sperm cell
   e. corona radiate

8. The sperm cells are attracted to the oocyte by a process called?
   a. meiosis
   b. mitosis
   c. fertilization
   d. capacitation
   e. chemotaxis

9. Sperm and oocyte unite through what process?
   a. meiosis
   b. mitosis
   c. fertilization
   d. capacitation
   e. chemotaxis

10. The cell resulting from the union of the sperm and oocyte is?
    a. corona radiate
    b. zona pellucida
    c. zygote
    d. morula
    e. blastocyst

11. The oocyte is surrounded by a protective barrier of cells called the?
a. corona radiate
b. zona pellucida
c. zygote
d. morula
e. blastocyst

12. The germinal period covers what weeks of development?
a. first 2
b. weeks 2-8
c. after 8-birth
d. all of the above
e. none of the above

13. Weeks 2-8 are what period?
a. fetal
b. germinal
c. embryonic

14. How long after fertilization does implantation occur?
a. 1 hour
b. 1 day
c. 3 days
d. 4 days
e. 7 days

15. The placenta and chorion will develop from the?
a. yolk sac
b. amniotic cavity
c. trophoblast
d. villi
e. morula

16. The organ responsible for the swapping of materials between the mother and fetus is the?
a. yolk sac
b. trophoblast

c. morula
d. placenta
e. endoderm

17. The layer which invades the uterine wall is the?
a. corona radiate
b. zona pellucida
c. syncytiotrophoblast
d. morula
e. blastocyst

18. The linings of the lungs and GI tract will develop from which germ layer?
a. endoderm
b. mesoderm
c. ectoderm

19. Skeletal muscles and bones will develop from which germ layer?
a. endoderm
b. mesoderm
c. ectoderm

20. The epidermis will develop from which germ layer?
a. endoderm
b. mesoderm
c. ectoderm

21. Skeletal muscles will develop from immature cells called?
a. myoblasts
b. myoclasts
c. myocytes
d. myocardium
e. myofilaments

22. The upper and lower limbs will begin to develop at the beginning of what week?

a. 2
b. 4
c. 6
d. 8
e. 10

23. Before birth there is a hole between the two atria called the?
a. fossa ovalis
b. foramen ovale
c. ductus arteriosis
d. ligamentum arteriosis

24. The second stage is labor involves?
a. the expulsion of the placenta
b. regular uterine contractions until the cervix is fully dilated
c. the baby exiting the body

25. Premature babies will most likely have trouble with what system?
a. digestive
b. endocrine
c. lymphatic
d. respiratory
e. urinary

26. What is needed to reduce the attraction of water molecules in the alveoli?
a. lipids
b. epithelial cells
c. oxygen
d. carbon dioxide
e. surfactant

27. Twins resulting from one fertilization are?
a. monozygotic
b. dizygotic

Answers to multiple choice questions

1. B
2. B
3. A
4. A
5. D
6. E
7. C
8. E
9. C
10. C
11. A
12. A
13. C
14. E
15. C
16. D
17. C
18. A
19. B
20. C
21. A
22. B
23. B
24. C
25. D
26. E
27. A

APPENDIX

Most all texts for Human Anatomy and Physiology have the exact same material in them. The better ones have thousands of illustrations and examples. You can choose the one you prefer or the one your instructor has selected. Below is a list of texts used as a reference.

Marieb, Elaine and Katja Hoehn. Human Anatomy and Physiology. Benjamin-Cummings Pub Co; (May 30, 2006).

Martini, Frederic; William C. Ober, Claire W. Garrison, Kathleen Welch and Ralph T. Hutchings. Fundamentals of Anatomy and Physiology. 5th ed. Prentice Hall College Division. January, 2001.

McKinley, Michael; Valerie O'Loughlin and Theresa Bidle. Anatomy and Physiology. 1ed McGraw Hill Science. January 6, 2012

Patton, Kevin T. and Gary A. Thibodeau. Anatomy & Physiology. 7th ed. Mosby. (February 26, 2009).

Saladin, Kenneth S. Anatomy and Physiology. McGraw Hill Higher Education; 5th edition (February 15, 2009)

Seeley, Rod R., Trent D. Stephens, and Philip Tate. Anatomy & Physiology. 9th ed. Boston, Mass. McGraw-Hill, 2010.

Shier, David; Ricki Lewis and Jackie Butler. Holes Human Anatomy and Physiology. 9th ed. McGraw-Hill. 2009.

Tortora, Gerard J. and Bryan Derrickson. Introduction to the Human Body. 9th ed. John Wiley and Sons, Inc. 2012.

Made in United States
Troutdale, OR
02/20/2025

29146433R00166